FRANCIS GALTON & ALPHONSE DE CANDOLLE

Francis Galton & Alphonse de Candolle

Notes & Correspondence

Edited by Gavan Tredoux

galton.org

https://galton.org

ISBN: 978-1-963225-10-5

Francis Galton: a Lifetime of Exploration.

1. *Francis Galton's Nature and Nurture:* 1822–1865.
2. *Francis Galton's Genius:* 1865–1911.

Supplements:

1. *Sir Francis Sacheverel Darwin:* Charles Darwin's Adventurous Uncle.
2. *Francis Galton at Cambridge:* Letters and Diaries, 1840–1844.
3. *The Diary of Charles John Andersson:* 1850–1851.
4. *Francis Galton on Mars:* the Discontinuous Variation Notebook.
5. *Francis Galton and Alphonse de Candolle:* Notes and Correspondence.
6. *Francis Galton and Alfred Binet:* Correspondence.
7. *Francis Galton:* Selected Papers, Notebooks and Diaries.

Contents

of Candolle's French texts, along with Galton's frank private notes derived from his reading of Candolle, and the full text of the review he published of the *Histoire des sciences.*

Chapter 1

Introduction

> Men of science sometimes lack moral strength and the result is a willingness to hide certain opinions rather than lie.
>
> —Alphonse de Candolle.[1]

Alphonse Louis Pierre Pyramus de Candolle[2] was born in Paris on October 27, 1806. He was a son of the noted botanist Augustin Pyramus de Candolle (1778–1841), whose Protestant family had originally fled Provence for Switzerland in the 16th century. Augustin was then based at Montpelier in France, but relocated back to his own birthplace, Geneva, ten years later, to take up the chair in Botany there. Alphonse actually studied law in Geneva, after first taking a BSc. degree. Though he qualified in 1829 as a Doctor of Laws, he did not practice law for long and instead joined in his father's botanical work. This

[1] Candolle to Galton, 1888/05/28.

[2] Referred to below as 'Candolle', rather than 'De Candolle', though note that Galton routinely used the latter, as did Darwin.

became a tradition. His son and collaborator Anne Casimir Pyramus (1836–1918) and his grandson Augustin (1868–1920) carried on the botanical family practice in Geneva to the fourth generation and beyond.

After marrying in 1832, Alphonse succeeded his father as Professor of Botany at the University of Geneva in 1835, holding the chair, while supervising the botanical gardens there, until his early retirement in 1850. Private means allowed him to pursue science for its own sake, and to dabble in politics, where he enthusiastically promoted the use of referenda. Physically he was 'tall in stature, with a prominent nose, and small, rather deep-set eyes'.[3] He had the family's peculiar ability to wiggle the scalp on demand, as he once gravely demonstrated to a curious friend.[4] His personality matched his 'distinguished' appearance: 'his manners though reserved, were always exquisitely urbane. If he lacked enthusiasm of a demonstrative sort he made up for it by extreme sobriety of judgment and inexorable persistence'.[5]

At first Candolle dedicated himself to completing his father's ambitious task of cataloguing all known plants in his *Prodromus*, eventually extending the project to 17 volumes before giving it up. Though he was elected to the French

[3] W. G. Farlow, 'Alphonse de Candolle', *Proceedings of the American Academy of Arts and Sciences*, May 1893, 406–411.

[4] Candolle to Darwin, 1871/03/08, DARWIN, DCP-LETT-7557.

[5] W. T. Thiselton-Dyer, 'Alphonse de Candolle', *Nature*, July 20, 1893, 269–271.

Academy of Sciences in 1851, his own distinct reputation was really made on the publication of his *Géographie botanique raisonée ou exposition des faits principaux et des lois concernant la distribution géographique des plantes de l'époque actuelle* (Rational botanical geography or exposition of the main facts and laws concerning the geographical distribution of the plants of the present time) in 1855.[6] Restricted by caution and an unwillingness to speculate inductively, he drew no direct evolutionary conclusions in this work. Nevertheless, he made an irrefutable case for the ancient origins of the distribution of plants, and, implicitly, for the impermanence of species.

Candolle was, in any event, a slow and reluctant convert to the theory of natural selection. 'The general hypothesis of indefinite transmission across centuries of forms with more or less marked modifications seems preferable to any other, but I am uncertain that natural selection is the means for it. There are so many factors that for a long time keep forms the same from generation to generation or that cause them to revert! It is so rare for a new form to be preserved without the protection of man! I know of no proven instance of the latter case. There are some, probably, but none has been proved, as far as I know'.[7]

Over the years Candolle maintained an extensive corre-

[6] Candolle 1855.

[7] Candolle to Charles Darwin, 1862/06/13, DARWIN DCP-LETT-3603.

spondence with Darwin, who he had first met as early as May of 1839, and ruefully venerated after the publication of the *Origin*. He did not embark on any foreign expeditions, working instead from the comfort of Geneva—wintering in the family mansion near the cathedral, summering in the country suburbs at Vallon—through a network of international contacts, much as Darwin himself had done once the momentous *Beagle* journey was over. In 1869 he was elected a foreign fellow of the Royal Society, as his father had been in 1822.

Candolle had been actively lobbying for election to the Royal Society for some time prior to 1869, annoying the Kew botanist J. D. Hooker, who wrote to Darwin as early as 1859 to complain that 'Alphonse De Candolle has written asking me to help him to election as Foreign fellow of R.S. & referring me to you, Murchison & Lyell. I am greatly shocked, having always thought him too much of a gentleman, though I never gave him credit for overmuch modesty'. Darwin recoiled. 'I am disgusted at A. De Candolle. How a man can think a begged honour, worth having I cannot understand!'[8]

After the publication in 1869 of Galton's *Hereditary Genius*, or perhaps at some time before that, as he later maintained, Candolle was stimulated to conduct his own examination of those causes which produce scientific eminence, *Histoire des sciences et des savants depuis deux siècles* (His-

[8] J.D. Hooker to Darwin, 1859/01/25, DARWIN, DCP-LETT-2404. Darwin to Hooker, 1859/01/28, DCP-LETT-2406. For Hooker see X Club. For Murchison, see Murchison, Sir Roderick.

tory of science and scholars over two centuries).[9] In Candolle's survey, he emphasized the advantages he saw in a broader European approach, confined to more recent figures.[10]

> Will our sources help to unravel what is due to heredity and subsequent causes? If that is possible. we will have a step beyond the curious research of Mr. Galton, in his recent work on the inheritance of [ability]. This author has gathered a lot of data on eminents of all categories: judges, men of state, scholars, writers, artists, etc., mainly of his country, and while speaking of education and like causes that push children into the direction of their fathers or ancestors, it highlights heredity. The very title and the first sentence of his book show that he sees it as the dominant cause ... I do not see, however, that he gave the evidence, nor that he scrutinized this question in any decisive way. It offers very great interest and piques curiosity, if only by its difficulty.
>
> My information about men of science has been collected in a different manner from that of Mr. Galton. I have employed more complete biographical documents, drawn from French, English and German books. I can flatter myself to have thus penetrated

9 Candolle 1873; Galton 1869. Although Candolle's book was nominally dated 1873, it was issued by November 1872 at the latest.
10 Candolle 1873.

> more deeply into the heart of the question. Mr. Galton studied, using biographical dictionaries, 65 of the leading scientists since Aristotle until our days. He found out who had fathers, brothers, sons or other close relatives, more or less eminent. I do not doubt in any way that choice, but confining my observations to the two centuries on which information abounds, by relying on a few hundred names instead of 65, and especially by using the membership lists of foreigners trained slowly and scrupulously by the three most competent scientific bodies we can find, I obviously have a broader base and reach than Mr. Galton's. As to the facts, we will agree completely. It is in the interpretation of the facts, in other words, in the search for cause, there will be a fairly large difference between our two opinions.

Galton's name appears many times in the text. Candolle worked hard to distinguish his conclusions from his predecessor, without denying the influence of heredity *per se*. The general drift of this was to claim greater degrees of nuance and sophistication, a variant of what might be called the 'argument from complexity', ostensibly examining the non-hereditarian factors which went to make up scientific men. Candolle argued that Galton had squinted past these. One of the botanist's principal conclusions—based on a statistical analysis of the family relationships of elected members of the principal scientific academies of Europe—was that *specific* talents for excellence in one field were not inherited, but rather a kind of *general* scientific ability or firepower, which might manifest itself, in the progeny of the eminent, in different scientific fields. This would be called 'general intelligence' today, and understood even more broadly, beyond the boundaries of science as such, but had already been explicitly argued for in *Hereditary Genius*.

The correspondence between the two men was initiated in late 1872, when the botanist sent Galton a complimentary copy of *Histoire des sciences et des savants*. Galton went on to review the book for the *Fortnightly Review* in March 1873, under the title 'On the causes which operate to create scientific men'.[11] The review is pointed but polite, and is reproduced in full in Chapter 3 below. Galton complained that 'my name is frequently referred to and used as a foil to set off his own conclusions. The author maintains that minute intellectual peculiarities do not go by descent, and that I have overstated the influence of heredity, since social causes, which he analyses in a most instructive manner, are much more important. This may or may not be the case; but I am anxious to point out that the author contradicts himself, and that expressions continually escape from his pen at variance with his general conclusions'. Here he was thinking mainly of race, which Candolle readily described as of primary

[11] Galton 1873b.

importance at the extremes of ability that his survey dealt with. Moreover Candolle insisted that mental qualities would be heritable in just the same way that physical ones were. 'Consequently, I propose to consider M. de Candolle as having been my ally against his will, notwithstanding all he may have said to the contrary'.

Galton's reading is strongly supported by a letter from Candolle to Charles Darwin a few years earlier. 'What a book we could write, either one of us, if we should wish to apply the ideas of naturalists and their methods of observation to the human species! Particularly to moral and intellectual facts. When one advances in age, one finds that one has accumulated documents about which one says nothing. I strongly believe, for example, in the heredity of moral disposition. I have known several families in which all the individuals, or almost all if you will, have been good, or almost all wicked, gay or sad. The exceptions are easily explained by the mothers'.[12] This is reinforced in another letter to Darwin, written after the publication of the *Histoire*. 'As a matter of fact, I do not think I have opposed Mr Galton, but, in succeeding him, I believe I have completed his work by drawing more varied and complete conclusions from similar facts. I hope that he will yield to this way of envisaging our work and indeed, he knows the extent to which I have rendered justice to his research'.[13]

---

[12] Candolle to Darwin, 1868/07/02, DARWIN, DCP-LETT-6264.
[13] Candolle to Darwin, 1873/01/14, DARWIN DCP-LETT-8737.

The extensive notes for Galton's review, prepared from his close reading and re-reading of the book, along with rather more frank commentary, are given below in Chapter 2.[14] 'I don't say that genius is the sole cause of success—quite the contrary. I do not see that he has [refuted me]. I have taken much higher names, selected with great care, & studied their histories from all kinds of sources. His men are so unknown to fame that he can learn hardly anything about them'.

Working assiduously through each page, Galton reacted acidly at times: 'most weak', 'Did it ever occur', 'stuff', 'stuff & nonsense', 'a concession', 'some crass error here', 'all gleaned from my book', 'concession after concession', and so on. He scorned at the idea that passing states of mind of the parents during the act of sexual intercourse could be inherited by the children. But parts of the book appealed strongly to him, since Candolle shared his antipathy to religion, especially in its unreformed Catholic state, and to classical education in Latin and Greek, which Galton actively detested. The extent to which religious dissenters—such as the Huguenots and other non-conformist migrants escaping persecution—dominated the scientific ranks traced by Candolle, neatly matched Galton's own comments on the topic in *Hereditary Genius.*

The Englishman also wrote back directly to the author to protest that, in the end—despite Candolle's disclaimers, qualifications and other attempts to differentiate himself—the

[14] See page 25.

botanist had simply made more or less the same argument as *Hereditary Genius*, properly understood. 'I must ... express no small surprise at the contrast between your judgement on my theories and your own conclusions. You say and imply that my views on hereditary genius are wrong and that you are going to correct them; well, I read on, and find to my astonishment that so far from correcting them you re-enunciate them'. 'I feel the injustice you have done to me strongly, and one reason that I did not write earlier was that I might first hear the independent verdict of some scientific man who had read both books. This I have now done, having seen Mr Darwin whose opinion confirms mine in every particular'. Still, he was quick to praise the book. 'One of the most striking things to me in your book is the chilling influence on scientific curiosity you prove to result from religious authority. The figures you give seem to me of the highest importance. I am also greatly impressed with the conditions of fortune (funds not land) and the desire for an European rather than a local reputation which you ascribe to religious and other refugees'.[15]

Candolle had also sent a copy of his book to Darwin in November of 1872. The naturalist wrote back from Down House with qualified enthusiasm. 'I began reading yr new book sooner than I intended, & when I once began, I cd not stop; & now you must allow me to thank you for the very great pleasure which it has given me. I have hardly ever read

[15] Galton to Candolle, 1872/12/27. See section 4.1, p. 77.

any thing more original & interesting than your treatment of the causes which favour the development of scientific men. The whole was quite new to me, & most curious. When I began yr essay I was afraid that you were going to attack the principle of inheritance in relation to mind; but I soon found myself fully content to follow you & accept your limitations'.[16] At the same time, the naturalist confided to J. D. Hooker that he was 'wonderfully interested in Decandolle's "Hist. des Sciences",—excepting the part about Man, which seems to me rather poor'. Nevertheless, the book 'exalts D.C. higher than ever in my estimation'.[17]

Despite their differences in emphasis, from the beginning Galton and Candolle rapidly found more to agree on than they did to dispute. Their early interactions resemble wrestlers half-grappling with each other, then ultimately thinking better of it. They shared a general commitment to liberalism and the idea of free inquiry, which was noticed by the *Liverpool Mercury* in Galton's *Fortnightly Review* piece.[18] And Galton had closed his first letter to the botanist generously. 'I feel, now that I have come to the end of this letter, that I have done little else than find fault, but I beg you to be assured that my general impression of the book is of another kind. I feel the great service you have done in writing it, and I shall do what I can to make it known, as it ought to be, in

[16] Darwin to Candolle, 1872/12/11, DARWIN DCP-LETT-8672.
[17] Darwin to Hooker, 1873/01/05, DARWIN, DCP-LETT-8726.
[18] *Liverpool Mercury*, Saturday 22 March 1873, 7.

England'. He also sent the botanist a copy of his article on 'Hereditary Improvement' published in January of the new year in *Fraser's Magazine.*[19]

Candolle had immediately sent a conciliatory note in reply. 'Unless it escapes me, in the 482 pages of my book a sentence or even a word that may cast doubt on my respect for your impartiality, your character and your talent for inquiry, can absolutely only be in error and contrary to my intentions. You have always sought the truth. I enjoyed your work a lot and if it was accepted practice to include extensive material from one author I would have quoted you more often'. But he gently disputed Galton's claims to priority, noting that he had first thought of his investigation in 1833, though he did not publish anything. 'My writing was very far along when I discovered your work. I read it with infinite pleasure, and I have just re-read the most important chapters'. He reiterated his interpretation of Galton's focus on heredity. 'You usually highlight, as the main cause, heredity. When you speak of other causes they are mentioned in passing without trying to unravel their particular importance, individually or in combination'.

Oblivious to the contradiction, which had struck Galton sharply, the botanist also reiterated his ideas about psychological states during coitus becoming hereditary, as acquired characters. 'I remember that at the time of the siege of Sev-

[19] Galton 1873a.

astopol there were people who had insomnia caused by the dreadful suffering reported by the newspapers. The events of 1870–71 caused a great deal of mental alienation and certainly disturbed many minds to a lesser degree. I would not be surprised if this was not a cause of increase of madness or idiocy in children born in 1871 in a part of Europe'. He acknowledged Galton's statistical criticisms about national variations in age cohorts, stating that he did not have those proportions, but offered the (rather feeble) idea that in countries with high birth rates 'the proportion of child deaths is all the greater as there are more children to be cared for, so that the number of adults per country is not as variable as one might think'.[20]

Galton responded magnanimously, without pressing harder on the weaknesses in the eminent Switzer's argument. 'I assure you I feel like yourself, that the subjects on which we differ are altogether subordinate to the common interest we have in arriving at the truth on the same line of inquiry'. 'Of the many topics in your work left unnoticed I regretted much not being able to speak of your most just criticism of the misuse of the word "Nature." For my part, I will never offend again unless through a slip of the pen'. (This was a rash resolution that Galton would not keep.) Still, he underscored the absurdity of Candolle's 'acquired experiences during coitus' hobby-horse (which the Switzer had shared enthusiastically with Darwin years earlier, but the naturalist never developed

[20] Candolle to Galton, 1873/01/03. See section 4.2, p. 82.

any enthusiasm for) with a pointed joke. 'A strong instance (if accurately recorded) of alcoholism combined both with the evil influences of close interbreeding and of old age on the part of one of the parents, in producing no bad effect on the offspring, is that of Lot and his daughters (Genesis xix. 31)'. Pragmatically, he proposed simple experiments on animals—rats on cannabis!—to establish if mental states in coitus could be inherited.[21]

Galton also carefully clarified his views on improvement of human stock. 'Encouragement of the best is the surest and safest way of discouraging the inferior. We are such a set of mongrels that except in extreme cases we should not be justified in "banning" any marriage. All we can say is, that some marriages are more hopeful than others. I therefore go no further at present, than urging that hopeful marriage should be encouraged'.[22]

The two men quickly made plans to meet in person. Galton was a frequent visitor to the Alps, and had produced extensive revisions of Murray's travel guides to the area in the 1860s.[23] However it was 1886 before they finally managed to bring it off. In the meantime they continued an active correspondence. Galton sent the Swiss a complimentary copy of his *English Men of Science* in 1874, and regularly dispatched copies of his journal articles, including his ground-breaking twin re-

[21] Candolle to Darwin, 1868/07/02, DARWIN DCP-LETT-6264.
[22] Galton to Candolle, 1873/05/07. See section 4.4, p. 105.
[23] Galton 1864, 1865.

search.[24] The extinction of families, first posed by Galton in 1873, was a problem Candolle had considered too, though he appears to have lacked the mathematical aptitude to follow the technical elaboration of the problem by Galton and Watson over the years.[25] The botanist was also greatly interested by the Englishman's inquiries, starting in the mid-1870s, into visualised numerals, mental imagery and the propensity to have visions, culminating in the publication of *Inquiries into Human Faculty* in 1883. So too his work on the inheritance of height and eye colour, summarized in *Natural Inheritance* in 1889, which expanded his interests into the mechanisms of heredity and what is now known as population genetics.[26]

Candolle kept up a stream of commentary, well represented here, on Galton's developing research, replying with copies of his own work. His *Origine des Plantes Cultivées* (1883), was translated into English in 1884 as *The Origin of Cultivated Plants*.[27] A second edition of his *Histoire* was published the same year (though neither edition has ever been translated into English). 'I am preparing a second edition of my History of Science and Scholarship, which has been out of print for a long time and which the booksellers request from me. For that I am making extensive use of your *English Men of Science*

[24] Galton 1874a, 1875a,b,c.

[25] Galton and Watson 1874. This is considered the foundation of the modern theory of branching stochastic processes.

[26] Galton 1883, 1889a.

[27] London: Kegan Paul & Co, 1884.

and the recent volume of Inquiries into human faculties which contains many curious articles. We practice the same method, that of observing, and when we can, counting to compare, therefore we must support each other and we stand to gain little in opposition'.[28]

A series of queries from Candolle about leading men of science Galton had known personally, including Charles Darwin and his accomplished sons, received thoughtful replies from the Englishman. 'He did not draw, he had not a good ear for music, but was much affected by it, sometimes to tears. He had naturally, (excuse the word which I know you detest! but I mean 'innately') a very emotional disposition, which was repressed by his habits of hard thinking, but always ready to burst out. Thus his delight in the scenery of a tour about the English Lakes a few years ago, had all the freshness and eagerness of that of a boy. However his nature could not be called aesthetic. As regards imagination I hardly know whether I understand the word in your sense, nor indeed if I have any definition of my own. I know that his faculty of mental imagery was once vivid and had become diminished, both from what he distinctly told me and from corroborative evidence. But that he ever was deceived by imagination I should think most unlikely, as he was so remarkably veracious'.[29]

Though both men maintained their differences in emphasis

[28] Candolle to Galton, 1884/01/14.
[29] Galton to Candolle, 1884/01/27.

and primary concern, Candolle undoubtedly shared Galton's meritocratic position. 'The true principle should be to obtain specific men: the right man in the right place. We are moving away from that in the current English Civil Service system. It is perhaps the consequence of the false ideas of the present democracy which judges all men equal and fit for everything, which leads to a general mediocrity'.[30] He also subscribed to the idea of natural abilities, and was well aware, as was Galton, that people evoke influences from their environment on the basis of those abilities and inclinations, which is now recognized as gene-environment correlation. 'My conclusion is that most men do the things they naturally feel comfortable with, and neglect the ones they are not good at. The use that results from a natural disposition, and the non-use, is also due to a natural weakness. This is contrary to what teachers and many parents think. They want to force young men and women to do what they do not like, while the youth would like to do what they want. There is thus a lot of lost time and energy; but the youth soon escapes the constraint and then we see the young people who are not calculators abandon mathematics, the young ladies close their pianos etc ... The pedagogues want to make all individuals alike, but the individuals prefer to be dissimilar, which will be a greater benefit to society in general'.[31]

[30] Candolle to Galton, 1889/09/15.
[31] Candolle to Galton, 1890/06/23.

Galton's copious and detailed notes from his reading of Candolle, salted with his commentary, are reproduced here for the first time. The thoughtful consideration by Galton of Candolle's *Histoire*, which appeared in the *Fortnightly Review* in 1873, is added for convenient reference. All the known letters between Galton and Candolle are also reproduced, translated to English where necessary, alongside the original French for comparison.

Many of the letters from this 'pleasant correspondence',[32] which ended only on Candolle's death at age 86 in 1893, have appeared in print before, but not in translated form, requiring reading knowledge of French. They originally appeared in Karl Pearson's second volume of the *Life, Letters and Labours of Francis Galton* and the second part of the third volume, at a time when fluency in French among the reading public was a more reasonable assumption than it is today.[33] Candolle's letters to Galton remain in the Galton Papers at UCL, but Galton's letters to Candolle appear to be in the possession of the Candolle family or their heirs; they had only been lent to Pearson, as shown by the following letter from the son of Alphonse, Anne Casimir Pyramus de Candolle (1836–1918).

Geneva December 13th 1911

Dear Sir

I have safely received the 14 letters of Francis

[32] Galton 1908, 291.
[33] Pearson 1924, 1930.

> Galton which I had lent to you, together with the copies of those of my father for which I thank you very much. I read them with a very great pleasure, thinking that I was again listening to him.
>
> You are quite at liberty to publish any extracts of these letters, you may think of sufficient interest to the public. Only I strongly advise you to have them printed from the originals, for the typed copies are indeed, as you say, far from satisfactory.
>
> Believe me, dear Prof. Pearson to be
>
> Yours very faithfully
>
> C de Candolle

The letters appeared again in an article by G. de Morsier published in 1972, 'Correspondance inédite entre Alphonse de Candolle (1806–1893) et Francis Galton (1822–1911)' (i.e. 'Unpublished correspondence ...') Morsier obtained the originals from Roger Alphonse de Candolle (1905–1989), Casimir's grandson, but, somewhat incredibly, seems to have been completely unaware that they had already been published by Pearson, with the exception of single brief note from Galton.[34]

Three previously unpublished letters have been added to those included by Pearson and De Morsier. Editorial commentary relates them to their context and clarifies obscurities.

All material is provided here for educational purposes only.

[34] Morsier 1972. There are many errors in this article's editorial commentary and bibliography.

Gavan Tredoux

November 2018.

Figure 1.1: Alphonse de Candolle as a young man.

Figure 1.2: Francis Galton at about 80 years of age circa 1902. Restored.

Figure 1.3: Alphonse de Candolle in later years.

Chapter 2

Notes on Candolle

The Fortnightly, 450 words to a page.

Alphonse De Candolle Histoire des sciences et des savants depuis deux siècles. Geneva. 1873.

3.[1] Naturalists have yet to learn that the infinitely small is as much a reality as the infinitely great.

4. All stationary condition is in all probability only so apparently.

7. Isolation probably as necessary as time, to consolidate new forms.

13. Man believes himself to be very susceptible of changes—

[1] The page number in the *Histoire*.

hence probability of modifications under selection. My (De C.!) researches prove ~~the exact~~ rather the contrary.

16. Edgar Quinet 'La Creation'. 2 remarkable volumes.

23. Royal Soc: London names 50 foreign members. Academie des Sciences Paris 8 associés strangers & from 40–70 correspondents. Royal Academy of Berlin —— foreign members.

29. Wishes to find a name for 'scientific discoverers' but cannot find one & uses 'savants' as a makeshift.

31. There is no complete list hitherto published of the French associés—mentions where they are to be found in separate lists. In 205 years there are 92 & 3 notables whom he does not reckon.

His authorities about their lives almost all biograph. dictionaries.

36. List of the 92 Associés with tabulated particulars.[2]

42. Their nationalities are the result of many researches, also those of the correspondents with following tables (plate) which are those of the years 1750, 1789, 1829 & 1869. He cannot

[2] Galton: He spells Bernouilli.

get a complete list to hand.

44. List of the 92 Associés & Correspondents, as above, 212 in all.

51. Royal Society of London. List of 1750 was 150 in number; that of 1789 was 96. He cuts these down to 72 & 65 respectively. In 1829 the maximum was already fixed at 50—. Total ———————— 235 in all.

61. Academy of Berlin—nominations at first irregular but at the beginning of this century they improved. In all cases foreigners are mixed with Germans. The latter have been eliminated in the table p. 64 for the above 4 periods, 195 in all.

74. Growing speciality of savants. In old times they were famous in more than one branch.[3]

75. Where there are no investments for many, like funds, the time of men is occupied with their own affairs. It is more easy to administer a property ~~in funds~~ 'mobiliére' of 500,000 fr. than one in land of 1,000,000 fr. much more easy in public securities.

[3] Galton: Murchison après avoir achevé honorablement une carrière militaire, est devenu un illustre géologue. [Murchison, after having completed an honorable military career, became an accomplished geologist. Candolle 1873, 75].

76. To end of last century the lists contained many jesuits, 'minimés' & abbés. Now there is the Abbé Haüy & the Jesuit Secchi[4] but the abbés are not parish priests—they had leisure & they have been suppressed by the Revolution.

78. Singular disproportion of Catholic ecclesiastics—also of missionaries either Catholic or Protestant.

79. The latter have never made natural history collections. Rome has nothing botanical to shew for her Propaganda, though the Jesuits were absolute masters of Brazil, Spanish America, Philippines & elsewhere. The Protestants have been a shorter time at work and have not been such masters, but they have had abundant leisure.[5]

80. The growing specialization makes it important to separate tuition from purely scientific work. Savants may lose in clearness of ideas by not having to lecture, but on the other hand they w^{d} not be tempted to sacrifice truth to clearness—facts being naturally complex.

82. Birth of savants not French noble 41 per c^{t} middle 52

---

4 René Just Haüy (1743–1822) the mineralogist; Pietro Angelo Secchi (1818–1878) the astronomer.

5 Galton: Hooker told me that the missionaries sent him nothing but that some overworked small country doctor, did.

labouring 7 yet the labourers are 66 to 75 percent of the pop: Birth of 36 French savants taken from Royal Society list gives the above divisions: noble 28 middle 47 labouring 25.

86. He makes a supplementary list of 24 of his own choice giving 46 33 21.

90. If natural talent & taste were the only causes that made a man a savant the labouring class would have been infinitely better represented.[6]

90. Poor men cannot be scientific—learning does not pay. It is a work of abnegation & cannot be recommended to the penniless.

92. The test of the disposition of a man, whether or no he is qualified to be an investigator is not examination but to see if he is curious about real facts & loves truth for herself, also he requires certain doses of perseverance & capacity "'l'individu doit vouloir et pouvoir."[7]

93. Do our documents enable us to disentangle what is due to heredity & to subsequent causes. If so, we shall have made

[6] Galton: This is a non-sequitur unless the capacity of all classes & their task is alike. I maintain, & so does de C: further on, that they are not alike. Exhibition, patronage, fellowship, professionalism.

[7] The individual must be willing and able.

a step in advance of Mr. Galton's curious researches. ... The title & first phrase of his book shew that he regards heredity as the dominant cause. Yet I do not see that he has proved this nor that he has scrutinized the question in a sufficiently special manner.

94. I have employed biographical documents more complete than those of Mr. Galton. I may flatter myself that I have penetrated more fully into the heart of the question. Mr. Galton has studied 65 of the principal savants from Aristotle to the present time following (d'après) the biographical dictionaries .... but I ... &c. We agree upon the facts but in the interpretation of the facts they will remark a great difference between our opinions.[8]

95. In botany from the first until 1851 there have been more than 5000 authors (table of Pritzel Leipzig 1851) & probably not as many in the other sciences, but say 2000 for each of the other 8 sciences, 16,000 authors in 2 centuries. Allow $\frac{3}{4}$-ter of them or 12,000 to be non-French & therefore eligible to be Associates, only 92 have been elected or only 7 or 8 per 1000. there are 4 cases of Father & son viz. 3 Bernoullis, 2 Eulers,

8 Galton: This is most inexact. I say genius is hereditary just like structure & this De Candolle admits. I don't say that genius is the sole cause of success—quite the contrary. I do not see that he has [refuted me]. I have taken much higher names, selected with great care, & studied their histories from all kinds of sources. His men are so unknown to fame that he can learn hardly anything about them.

2 Herschels.

98. It is impossible to know the number of sons of Associates who occupied themselves in some way in science, but among the correspondents are Caponè, Saussure, de Candolle, also other sons of associates now alive may yet be named.[9] He adds as deserving notice 3 more Bernoullis & the son of Linnaeus.

99. Gives very imperfect accounts (avowedly) of their other distinguished relations & these only by male line.[10] Finds more eminent descendants than ascendants & agrees with me.

101. From these facts he concludes in a sense, rather contrary to the actions of heredity properly so called & thinks it has little effect except in the mathematical sciences, but that education example advice preponderate, because (1) All young men have some college education if they differ it must be from family causes but these need not be only heredity the rest is not intelligible—but he says 'the descendants of men accustomed to work with their brain are more disposed to study than those of men who have only developed their muscles.' (2) A large proportion are sons of clergymen, 13 out of 90 associates only 5 are sons of medical people—now if capacity for the sciences were an affair of heredity there would be many more

[9] Galton: This points to their relative obscurity.
[10] Galton: Ought to have done this better.

sons of doctors on the list.[11]

104. (3) If heredity gave an instinctive aptitude the proportion of scientific men in the same family w^{d} be alike in all countries, but it is greater in Switzerland. Here education &c. counts for much.

105. Note. Has made a list of all Swiss attached to one or other of the 3 societies. They number 29, only 11 of these have no near relation known in science & of those 11, 3 or 4 has no surviving son. The other 18 have had 20 relations authors & in science. 11 cases for S; 5 brothers; 2 gr: to P.

106. It is exceptionally rare that a savant had a mad or idiotic father except the disease was the result of a malady or old age.[12] A savant must at least receive such & such moderate gifts from his parents, & they must be combined in a happy manner so as to raise hm above mediocrity.

107. Says that in this way a man endowed with a strong dose of perseverance, attention & judgement, without much deficiency in other faculties may succeed in various professions. He does not believe in innate & [...] vocations except probably for mathematics.[13]

[11] Galton: most weak.
[12] Galton: Did it ever occur.
[13] Galton: a concession.

109. Biographies of naturalists shew that they sometimes have a precocious & determined disposition (penchant) for observation.[14]

110. Compares mathematical & natural history aptitudes, the one for rigorous reason in a single line, the other attending to many none of wh: is rigorous.[15]

112. In admitting a certain degree of heredity of faculties there must be two sorts of heritage one appropriate to mathematical & the other to natural history studies.

112. You seldom see in the same family poets or artists & savants or literary men of a high order.[16]

113. Again distinguishes between general & special heredity.

114. Education—curiosity is the principal motive to discover, for real things. If you tell a child continually not to occupy himself above such & such things—& not to be curious you make him indifferent or timid.

115. Au point de vue moral, c'est une bonne habitude de chercher ce qui est vrai. Il en résulte plus de véracité chez la

[14] Galton: still more so.
[15] Galton: also arts.
[16] Galton: Mendelsohn. Herschel.

moyenne des individus.[17]

116. A chance word may determine a series of researches. So a book without pretense as Mrs. Marcet to Faraday.[18] Good teaching, what it consists in.

117. It is strange how often the mothers of illustrious men have been mediocre & how often the pupils of the most celebrated professors have been second rate.

118. Libri attaches weight to the causes which increase the general moral force of men.[19] In Germany the universities have long existed but it is only of late they have so progressed.

120. Religion, leaving aside the Greek Church, the non-French Roman Catholics are one & a half of the Protestant & yet they have furnished only $\frac{1}{4}$ as many associates, 25. Productive ratio = 1:6.

122. By Royal Soc: London lists the non English catholics are 4 times less productive then non-English Protestants. Comparing populations living side by side not one English is Irish

[17] From the moral point of view, it's a good habit to look for what is true. This results in more truthfulness for average individuals. Galton: very good.

[18] Faraday read Mrs. Jane Marcet's text *Conversations on Chemistry* (1806) as a boy.

[19] Guglielmo Libri Carucci dalla Sommaja (1803–1869), who wrote a history of science and mathematics in Italy.

Catholic on the French lists though they number $\frac{1}{5}$th as many as the Protestants. Austria is not represented and Catholic Germany is almost wholly deficient. In Switzerland the scientific productiveness of the Protestants is 26 times greater than that of the Catholics. Hence religion has a great effect.[20]

124. The principle of authority diminishes curiosity & augments timidity. This timidity must ~~ought~~ to become in some degree hereditary. As ~~terror of man~~ timidity is the case in animals & in man certain races classes & families are more courageous than others.[21] So a population educated in many generations under the principle of authority must become naturally more timid in intellectual matters.

125. A large number of the savants are sons of Protestant ministers—fortunately for science their fathers were not compelled to be celibate. If you cut off from the list of savants the sons of clergymen an equity will almost be re-established[22] between the scientific producing power of the two creeds. Thus a rule of pure discipline has had serious effects on science. The numbers of persons able to rear a family with moral simple &

[20] Galton: Recollect the Huguenots &c.

[21] Galton: concession, concession, concession.

[22] Galton: There must be some crass error here, see p. 40. Since 1827 are given professions of 21 parents of associates since 1827. Only 1 of these associates was Catholic & of the 20 parents of the rest only 4 were clergy & 2 Quakers (Young & Dalton) leaving 14 not clergy (at least).

laborious habits is small & cannot bear to be reduced.

127. Geneva shews the evil effect of religious restraint from 1535 to 1725 the reformers reigned, everybody was educated but no Genevan during all that time was distinguished in science. 1720–1730 the Calvinist powers diminished. In 1739 one was elected associate & ever since then Geneva has produced scientific men in a remarkable proportion.[23]

127.[24] Family traditions—some du *far niente*[25] some for doing nothing for nothing—some that it is right to labour for good without reward. Also standards for imitation.

129. Many political refugees implored their descendants to avoid politics.

130. Weiss. Histoire des réfugiés protestants de France depuis la révocation de lédit de Nantes, jusquà nos jours. 2 vols 8vo Paris, 1853.

131. Table of scientific descendants of Protestants expelled. There were 44 of whom 35 were in Switzerland.

---

[23] Galton: But these were sons & descendants of refugees—when did the parents arrive?

[24] 128.

[25] Italian: do nothing.

132. He supposes the total emigrant population to be 1 million of which the French part was half. This supplied 4 foreign associates 29 to a million. All the other non-French Christian populations are probably 150 to 300 millions & ought at same rate to have supplied 1600 associates, but they actually gave 88 or one eighteenth as many.[26]

135. in note, the refugee population is not likely to have much increased because it consisted chiefly of persons of a middle or superior race with few proletaires.

136. The refugees in other countries have produced fewer savants but more men distinguished in other ways. In Geneva they became distinguished for science chiefly in the 3^d^ or 4th generation showing the effect of the place &c. in directing their studies.

138. The refugee traditions were: not to be idle, to work with a will & for the public, to avoid politics, & to look for approbation to the enlightened men of all countries. Now local families of the same energy would rather look to local reputation. Also their property would be in land which requires overlooking. The refugee families have more leisure. The refugees of the XVI century have contributed most to

[26] Galton: Remark this. It shows what might be done by caste intermarriages.

science in proportion to their numbers. Thus the Huguenots, were enlightened gentlemen of letters for Protestantism had then great hold in there. Those after the revoked Edict of Nantes were chiefly merchants, industrials or agriculturalists & their descendants were like them.

140. J. Jacques Rousseau descended from a Huguenot bourgeois of Paris who became citizen of Geneva in 1555. (X) Note: this origin is given in Galiffe Généalogies genevoises 4vol 8vo.

141. The English emigrant puritans resembled in disposition the Huguenots. They gave Franklin & Rumford.[27] There have been more political Polish refugees, during a century these have produced no scientific descendants.[28]

141 continued. So the enormous emigration to the United States gives small intellectual results. It is composed of hard working men—sometimes run away with political ideas. Little favourable to the sedentary & unremonstrated labour of science. A great part comes from Ireland, a most unscientific country.

144. Influence of opinion. He notes 6 tendencies to 1. acquisi-

[27] Sir Benjamin Thompson, Count Rumford, FRS (1753–1814).
[28] Galton: [...] the same as to French emigrant noblesse.

tion 2. pleasure; 3. politics; 4. religion; 5. discovery of truth; 6. arts.

145. Can religion & science march together. He ~~doubts~~ thinks that they can, although in science a man only seeks truth regardless of consequences, while a man attached to a particular form of religion is persuaded that he has the truth & does not like it to be discussed. It is repugnant to him to hear certain deductions criticised.[29]

146. The scientific man weighs evidence & holds by the balance of evidence & is always ready to abandon his position when a new one seems better. But both are alike in pursuing ideas in preference to material interests.

147. Religion, art & science occupy one side of the moral sphere of man & love of self the other. Governments generally confound too much teaching with scientific progress. Science lives in liberty & individual labour.

148. Note speaks of complaints of professors having time occupied in examining.

149. Sovereigns ought to leave academicians leisure; constitutional governments hardly have the power to do so, the public

[29] Galton: This is very good.

won't understand.

150. Democracies are helpful in this way, for when they persecute it is not too long, their majorities being variable. They also do not attract the best men to politics.

152. The principle of aristocracy is individual speciality, that of democracy is to consider all persons as equally good for everything. It is hurtful to true specialities but raises the average of the nation.

153. A small country has advantages over a large one in not offering high political prizes. So their best men aspire to a European fame. Also they avowedly cannot suffice to educate their youth who without loss of amour propre travel to the best universities in Europe.

154. Importance of a language in wide use.

156. Geography. No associate has been born south of the Pyrenees, nor of Central Italy.

157. Inconveniences of a hot country to naturalists.

160. Classification of the 92 associates by nationalities, when in the XIXth century Germany 17, England 14, Denmark 1,

United States 1, Italy 4, Sweden 1, Switzerland 2.
160. Hence the population of a country is a very secondary element to the production of the highest order of scientific men. To succeed il faut vouloir et pouvoir.[30]

165. The small protestant countries offer the most remarkable proportions Switzerland, Scotland.[31]
167. The small towns are usually their birth places, except Paris.

169. Italy may be considered as a whole for there was greater uniformity of culture than in the separate states of Germany or cantons of Switz: where the savants of one never go to live in another.

170 Comparison of 'associates' by nationalities with the joint 'associates and correspondents' also by nationalities. He thinks their difference worth record & [speculates]. It shows the English to be more original because $\frac{A}{A+C}$ is greater with them than with the Germans. England $\frac{294}{226}$, Holland $\frac{65}{83}$ Germany $\frac{25}{25}$. He says Englishmen are not Professors & have more leisure besides their character being more original. (note) has tried

[30] You have to be both willing and able.
[31] Galton: Switzerland has greatly gone off since the previous period, men out of its total 12 associates (see p 161) only 3, Haller & Euler & De Cronsag are pure Swiss de Candolle counts 4, that is De la Rion, but he says p 140 that he descends by mother from Huguenots & is [Fr.]

in vain to separate associate professors from non-professors.[32]

172. Table. Nationalities & epochs.

184. Table XI on the scientific value of a million inhabitants of various countries.

188. Table XII In which the order of scientific value of the various nations is given. Switzerland has an immense superiority.[33]

189. Holland is low on the list yet her pop^{n} is remarkably instructed. It is not knowledge which has weakened but her desire to seek for truth without pecuniary advantage.

190. The small pop^{n} of Geneva 35,000 had 0.050 of the asso-

[32] Galton: I do not. The numbers vary no more than other statistical results from small numbers.

[33] Galton: His tables should be recalculated. There is an error in this in taking the pop: generally because increasing pop^{s} have an undue proportion of children & therefore shew at a disadvantage. It would be better to take the associates & correspondents of 1869 as compared with the pop: of 1829, because the children of the former date would have become men & would afford the stock out of which the selection was made. In this case France & England would have the respective values of 0.012 and 0.011 instead of the tabular values 0.010 & 0.008. But this is still unjust to England whose infant mortality is great & whose pop^{n} therefore includes more children than France Also the large pop^{n} of Ireland see p 190 is included & Ireland supplies nothing, allowing $5\frac{1}{2}$ Catholics the [case] for the remaining 18 millions would be 0.014). The true comparison would be between the nations to the pop^{n} above the age say of 45.

ciates & sometimes 0.100 of the Assoc: & Corresp: together whence the scientific value of a million as in table XI w^{d} be fabulous. ([...] It would be 29,000 as against 0.035 for Switz generally & 0.009 for Germany, or 3000 times as much).

193. Advantages of small towns & countries to science. They are all frontier & inhabitants must compare their own situation &c. with those of their neighbours, & emigration to escape tyranny is easy.

195. Race. Europeans & their descendants are the only people who play a part in the sciences. Race stands in advance of all other conditions. This is to be understood throughout.[34]

196. List of favourable circumstances. 1. [Competence][35]; 2. lettered class; 3. national traditions; 4. immigrants; 5. family traditions; 6. elementary education; 9. educational appliances; 8. public curiosity; 9 liberty of publication; 10. social considerations; 11. liberty of profession & residence; 12. tolerant church; 13. clergy favors science; 14. clergy married; 15. a widely spread language; 16. small country; 17. climate; 18 civilized neighbours.

[34] Galton: Quote the original. A complete concession.

[35] Candolle has 'Considerable proportion of people belonging to the rich or well-to-do classes of the population, relative to those who are forced to work constantly to make a living and especially to work at manual labour.'

198. + used when favourable, – when unfavourable.

200. The favourable conditions in Switzerland.

202. The protestant & catholic cantons.

204. Extreme tolerance to religious profession in 1735.

207. Has a gird at Bâle & at the German cantons for money seeking. Democracies are repugnant to gratuitous service and to established positions.

208. The Helvetic Society shows that out of $2\frac{1}{2}$ million of population about 1,600 persons are more or less disposed to make scientific researches. These are chiefly in the Protestant cantons.

211. Singular analogies between Bâle & Holland. Too severe an introduction prevents person following out new things. To know much is the contrary to absorbing oneself i the speciality.

214. France, the spoliation of benefices.

215. No French ecclesiastics whatever on the lists of non-French societies. The principle of authority dominates more

than ever in the Church.

216. The well to do class has augmented largely but the people love pleasure & faction much more than study & facts. The journals & their false news & feuilletons—different to German papers who hen debarred from politics, give facts.

217. Yet Germany at the end of last century preferred fictions.

218. Paris has a selected population—out of 64 savants of high distinction 16 were born there & 48 elsewhere. note. Fortunately the most dangerous part of a pop^{n} of a large town is that which leaves fewest descendants (illegitimacy) 'Selection' is the principal cause of the superiority of Parisians.[36]

221. England.

224. Irishmen shew more disposition to the work of the imagination than to practise research.

225. Scotland. Hopes a good future for England. Regrets the growing tendency to ask Government aid. Ought to study the soil effects in other countries.

226. Germany. Her scientific advance is new, & dates from

[36] Galton: Again for heredity.

1830-40. The cause must have begun to act yet earlier, say 1820–40 or 50, & we must compare Germany of those dates with her neighbours &c.

228. In XIXth century accession of Jews to the learned classes. Educational movement & change of opinion favourable to positive facts.

229. Her universities had always been the same, some even for 300 or 400 years. Oken's lectures about creation in 1827 at Munich were nonsense.

230. Profound & general changes of opinion are generally difficult to explain. Except at that after revolutions men desire peace. Hobbies weary. But they usually result from the reflexions of those who were children when certain great events occurred. The present generation of Frenchmen has been formed by theirs: Victor Hugo & Alexandre Dumas. The youths of today are formed under very different sentiments.

231. Italy. Genius has never been wanting.

233. The rich class have always been distinguished for culture & have yielded many savants. Unfortunately now there is a great commercial deviation.

235. United States. New England was very favourably circumstanced. There is now danger from immigration of men of a very different stamp & poor emigration of the old stock.

236. Americans are very capable of sacrifice for an idea. they are over curious about imaginary things. Religious sects, spiritualism.

237. Traditions favourable to steady scientific pursuits are absent.

238. Poland gave Copernicus but her emigration has yielded nothing.

240. Russians are much occupied with landed estates & military service but are ready to work gratuitously & the women are zealous for instruction.

241. Belgium had loss from expelled Protestants that she has not recovered.[37]

243. Spain & Portugal, quotes Galton.

244. Fears much that Galton figures are not exact.

[37] Galton: See "Office, Holy" Penny Cycl: for authorities—(Science) a very able & elaborate article.

245. The Arabs notwithstanding what men say, had only a pale reflexion from the degenerate schools of antiquity.

247. Final reflexions.

248. Importance of race.

249. Evidently many of the influences p. 196 are connected. This adds to their importance.

251. Draw on a map of Europe the principal events of modern history & we shall see how necessary the causes favourable to science have accumulated in a triangular space between middle Italy, Scotland & Sweden. These are the Renaissance—the Reformation & political liberty in England.[38]

252. Again about the triangle—or sector.

253. Often illustrious men blaze forth, just as the social state which had nursed them breaks to pieces; but they were the result of influences many years apart.

253. Résumé: the first causes are 1. European Race accustomed for generations [at least] in some classes, to intellectual

[38] Galton: This sh' be drawn.

labours.[39] 2. Climate

254. Not too hot; 3. adjacent to centres of intellectual culture; 4. habits of liberty.[40]

260. Germany after 1815 for the first time, had not (7 years – 30 years wars) circumstances may turn almost the whole of the men in a nation who are capable of a scientific career, away from it.

262. Speaks of an hereditary tendency to intellectual occupations.

263. List of recent grand discoveries & the countries that produced 62 discoverers. Spectral analysis; transformation of forces; ancient extension of glaciers; antiquity of man; natural selection; alternate generation. These give evidence [&] corroboration of all the rest.

266. Scientific work is much more collective, than it may appear.

267. Average capacity of a nation.

[39] Galton: Race.
[40] Galton: ? as to ancient Greece.

268. It is the fashion to look on a nation of unequal capacity & on families as equal in the same popn, but if one reflects one soon sees that the causes of inequality act with a still greater force on the latter & that the differences must be very sensible.[41]

270. Great poetry & activity have ordinarily preceded scientific men. He agrees with Galton but begs his readers to give more weight than he has done to surrounding circumstances.

271. The great scientific epochs coincide with a development of moral & scientific ideas. He unfortunately cannot find a criterion of eminence in the latter as so much depends on the form in which they are put. Almost every [...] idea exists.

275. The literary man has to use what exists. Also the personal position of the writer greatly enhances the effect of his writings & gives reputation.
277. Results from French Academy.

280. There are decided resemblances between the two categories. Hence they go together.[42]

281. History of the savants has not appeared so favourable to

[41] Galton: all gleaned from my book. Quote this strong concession to Hered: genius.
[42] Galton: one does not rob the other; concession after concession.

effects of heredity as Galton states.[43]

282. 'État momentairé lors de la conception' &c.

285. Tendency of mixed populations to form sub races thus German Jews are extraordinarily musical. Material likeness does not go with intellectual tendencies.

287. On teaching in schools. For numbers of years boys are wholly taught abstract or internal things.

289. Drawing of real objects should be taught.

296. Dominant language, estimate of pop^{n} 100 years hence.

	millions	at present millions
English England 124 million. US & colonies 736 million.	860	77
German	124	62
French	$69\frac{1}{2}$	$40\frac{1}{2}$

298. Hence Germany & France will be to England what Holland and Sweden now are to themselves.[44] Works published or translated into English will have an enormous circulation, so all outlandish books be translated into English, that German

[43] Galton: stuff & nonsense
[44] Galton: quote from this.

French & all the rest may read them.

299. The relative ease of using different languages for everyday purposes—Latin is topsy turvy—translate an ode of Horace word for word & it gives the idea of a building whose door is on the third floor. It is no longer a possible language. Civilization now requires short words & phrases.

301. English is much the most practical except in its absurd orthography.

302. In families where 2 languages are spoken, French always beats German.

303 English beats French.

304 Responsibility of English authors to keep the language pure.

307. Have a joint commission English, US authorities to amend orthography.

309. The laws of balance in the organs & their function.

310. Quotes Büchner traduction francaise approuve par l'auteus, à Paris chez Rhein wald.

311. Again l'état temporaire des parents un moment de la conception.[45]

313. For observers of heredity elderly people are wanted who recollect parents at same age. They should be good observers.[46]

324. Breeders have proved that in many sorts of domestic animals certain breeds (lignées) are more intelligent than others.

326. In human physique, heredity is incontestable now moral & intellectual phenomena are certainly bound up with physical in many if not all cases, hence moral & intellectual manifestations are in some degree under the influence of heredity.[47]

330. It is very curious to compare a school of mixed German & Italian children—they differ more than grown up people.

331. Again about the moment of conception—quotes from Lucas—(332) L'Enfant du Jubilé.[48]

333. On bastards.

[45] Galton: Stuff.
[46] Galton: ask if there are foundling hospitals abroad for investigation.
[47] Galton: Concession—to heredity.
[48] Galton: stuff.

337. It is reported that in the US the negro race has become more robust owing to selection in parents.[49]

343. In selection, adaptation does not mean selection.

344. Success is often due to improper means.

362. Thinks it impossible to rear fine races—no chance of continuity of effort.

368. In civilised countries there are categories of professions suitable to infirm individuals who would have perished in a savage society but who if they are intelligent honest & well educated thrive now.

372. Longevity & health—quotes le compte d'Angeville Essai sur la statistique de la population Francaise 1. vol 4to Paris 1836 with maps. Normandy & Brittany have equally few rejected conscripts yet longevity is much greater in Normandy.

380. Number of savants sons of pasteurs. He says I assert the contrary.[50]

[49] Galton: Q^y authority.
[50] Galton: he misrepresents me. I spoke of very pious people.

383. No sons of pasteurs are Generals.[51]

385. Ancient & modern opinion is that the poor are the great suppliers of population. Proletaire 'ad prolem generandum'.
388. Speculation on the inevitable extinction of family names, which he thinks original.[52]

391. H. Spencer's views that productiveness is antagonized by high development. This must be especially true as regards females.

392. Note brain often affected in Swiss girl students by overwork.

393. When a religion is once introduced into the poorer classes it extends rapidly owing to their proletarianism. Thus the Irish introduce catholicism.

404. Comparison of the Jews & the Christians (405) the practice of each is the religious profession of the others.

404. Mortality among children is smaller for Jews than Christians (Prussia).

[51] Galton: Nelson son of a clergyman. Clergymen cannot in England afford to put their sons in the army.
[52] Galton: Give working out at end of this MS.

409. L'Anglais est un animal politique (note) wishes he had had facts about Quakers.[53]

From Smiles Huguenots—Appendix.[54]

		England originally quite pastoral.
	1100	
		Flemings—Carlisle & border lands (name Fleming common in Westmorland—Cumb[d]) Then transported to South Wales. 'Gower' is still a distinct arm.
	1200	

[53] Galton: Offer to procure them for him.

[54] Samuel Smiles *The Huguenots, their settlements, churches, & industries in England and Ireland* (London: John Murray, 1869). Third Edition.

The 'cursede forrainers' were repeatedly attacked by the native workmen guilds [very] obstructive & troublesome		Flemings at Worsted near Norwich—hence worsted work—also Brighton—Newhaven. (Edward III was father of British commerce & introduced Flemings largely.)
	1300	

Wat Tyler, the Flemings were the first he attacked.		War between Eng. & France blocked out English wool from Flemish markets. More Flemings came over & were protected by Charter, p. 454 quotation from Michelet that the character of the English was seriously modified by them. Brothers "Blanket" made blankets & ? got their name from their Dutch clock maker. German names. Silk weavers from Italy—armourers. Queen Elizabeth too. First English saw mills—paper mills & wire mills were by Dutch & Germans. Needle makers felt hats by Spaniards & Dutchmen. Glass by Venetians.
	1400	

		Foreigners were the artisans in all that required special skill
	1500	
	1600	
	1700	
	1800	

Southerden Burn 1846. History of the French Protestant refugees settled in England. & other works, see p. 496.[55]

[55] John Southerden Burn *The History of the French, Walloon, Dutch and Other Foreign Protestant Refugees Settled in England from the Reign of Henry VIII to the Revocation of the Edict of Nantes* (London: Longman, 1846).

Chapter 3

On the causes which operate to create scientific men.

On more than one occasion I have maintained that intellectual ability is transmitted by inheritance; and in a memoir published last year in the "Proceedings of the Royal Society" I endeavoured to explain what ought to be understood by that word "inheritance."[1] Two points were especially urged—the first, that each personality originates in a small selection out of a large batch of wonderfully varied elements, which were all latent and competing; and secondly, that these batches, and not the persons derived from them, form the principal successive stages in the line of direct descent. Hence follows the paradoxical conclusion, that the child must not be looked upon as directly descended from his own parents. His true relation to them is both circuitous and complicated, but ad-

[1] Ed: *Fortnightly Review*, March 1873. See Galton 1872, 1873b.

mits of being easily expressed by an illustration. Suppose an independent nation, A, to have been formed by colonists from two other similarly constituted nations, B and C; then the relation borne by the *representative government* of A to that of B and of C is approximately similar to what I suppose to be the relation of a child to each of his parents. But the existence of a slender strain of direct descent is shown by the fact of acquired habits being occasionally transmitted. We must therefore amend our simile by supposing the members of the governments of B and C to have the privilege of making emigration easy and profitable to their constituents, and also, perhaps, the governments themselves to have the power of nominating a few individuals to seats in the Legislative Council of A.

It appears to me of the highest importance, in discussing heredity, to bear the character of this devious and imperfect connection distinctly in mind. It shows what results we may and may not expect. For instance, if B and C contain a large variety of social elements, it would be impossible, without a very accurate knowledge of them and of the conditions of selection, to predict the characters of their future governments. Still less would it be possible to predict that of A. But if the social elements of B and C were alike, and in each case simple, such as might be found in pastoral tribes, then the character of their governments and that of A could be predicted with some certainty. The former supposition illustrates what must occur when the breed of the parents is mongrel; the latter, when it is pure. Now, no wild or domestic animal is so mongrel as man, especially as regards his mental faculties; therefore, we cannot expect to find an invariable resemblance between the faculties of children and those of their parents. All that could be expected on the hypothesis of strict inheritance we do find; that is, occasional startling resemblances, and much more frequently partial ones. From this we have a right to argue that if the breed of men were more pure, the intellectual resemblance of child to parent would be as strict as in the forms of the equally pure breeds of our domestic animals.

I propose to refer in this article to a volume written by M. de Candolle,[2] son of the late famous botanist, and himself a botanist and scientific man of high reputation, in which my name is frequently referred to and used as a foil to set off his own conclusions. The author maintains that minute intellectual peculiarities do not go by descent, and that I have overstated the influence of heredity, since social causes, which he analyses in a most instructive manner, are much more important. This may or may not be the case; but I am anxious to point out that the author contradicts himself, and that expressions continually escape from his pen at variance with his general conclusions. Thus he allows (p. 195) that in the production of men of the highest scientific rank, the

[2] HISTOIRE DES SCIENCES ET DES SAVANTS DEPUIS DEUX SIECLES. Par ALPHONSE DE CANDOLLE (Membre Corr, de lAcad. Sciences, Paris; Foreign Member, Royal Soc., etc.). Geneva, 1873.

influence of race is superior to all others (prime les autres en importance); that (p. 268) there is a yet greater difference between families of the same race than between the races themselves; and that (p. 326) since most, and probably all, mental qualities are connected with structure, and as the latter is certainly inherited, the former must be so as well. Consequently, I propose to consider M. de Candolle as having been my ally against his will, notwithstanding all he may have said to the contrary.

The most valuable part of his investigation is this: What are the social conditions most likely to produce scientific investigators, irrespective of natural ability, and, *a fortiori*, irrespective of theories of heredity? This is, necessarily, a one-sided inquiry, just as an inquiry would be that treated of natural gifts alone. But for all that, it admits of being complete in itself, because it is based on statistics which afford well-known means of disentangling the effect of one out of many groups of contemporaneous influences. The author, however, continually trespasses on hereditary questions, without, as it appears to me, any adequate basis of fact, since he has collected next to nothing about the relatives of the people upon whom all his statistics are founded. The book is also so unfortunately deficient in method, that the authors views on any point have to be sought for in passages variously scattered; but it is full of original and suggestive ideas, which deserve to have been somewhat more precisely thought out and much more compendiously stated.

Its scheme is to analyse the conditions of social and political life under which the principal men of science were severally living at the four epochs 1750, 1789, 1829, and 1869. The list of names upon which he depends is that of the foreign members of the three great scientific societies of Europe—namely, the French Academy, the Royal Society, and the Academy of Berlin—in each case about fifty in number. There is a yet stricter selection on the part of the foreign associates of the French Academy, who number only eight at a time, and of whom there have been only ninety-two[3] in the last two

[3] List of the ninety-two foreign associates of the French Academy (three names of no scientific importance having been omitted, who were elected in early days—these are: Lord Pembroke, 1710; Duc dEscalone, 1715; and Prince Loewenstein-Wertheim, 1756). The names are arranged in the order of their election, and a hyphen (—) divides those elected before and after the year 1800:—

Denmark:—None.—Oersted.
England:—Newton, Sloane (Sir Hans), Halley, Folkes, Bradley, Hales, Macclesfield (Earl), Morton (Earl), Pringle, Hunter, Priestley, Banks, Black.—Maskelyne, Cavendish, [Jenner], Watt, Davy, Wollaston, Young, Dalton, Brown (Robert), Faraday, Brewster, Herschel (Sir John), Owen, Murchison.
Germany (Ancient Confederation):—Roemer, Leibnitz, Tchirnhausen (de), Wolff; Margraff, Herschel (Sir Wm.).—Pallas, Klaproth, Humboldt (de), Werner, Gauss, Olbers, Blumenbach, Buch (de), Bessel, Jacobi, Tiedemann, Mitscherlik, Lejeune-Dirichlet, Ehrenberg, Liebig, Wöhler, Kummer.
Holland:—Huyghens, Hartsoeker, Ruysch, Boerhaave, Van Swieten, Camper.—None.
Italy:—Guglielmini, Cassini (Dom), Viviani, Poli, Bianchini, Marsigli, Manfredi, Morgagni, Cervi, Poleni, La Grange (de).—Volta, Scarpa, Piazzi, Plana.
Poland:—Jablonowski.—None.
Russia:—Euler (the son).—None.
Sweden:—Linnaeus, Bergmann, Wargentin.—Berzelius.
Switzerland:—Bernoulli (Jacques), Bernoulli (Jean), De Crousas, Bernoulli (Daniel), Haller (de), Euler (Léonard), Tronchin, Bernoulli (Jean II), Bonnet (Charles), Sausure (Hor. Ben. de).—Candolle (Aug. Pyr. de), Rive (de la).
United States:—Franklin.—Rumford.

hundred years. It is remarkable that we find in this very select list four cases of father and son—namely, a Bernoulli and two of his sons, the two Eulers, and the two Herschels.

From an examination of these lists the author draws a large variety of interesting deductions. He traces the nationalities and the geographical distribution of the distinguished men of science, and compares the social conditions under which they lived. He finds them to be confined to a triangular slice of Europe, of which middle Italy forms the blunt apex, and a line connecting Sweden and Scotland forms the base; and then he shows that out of a list of eighteen different influences favourable to science, such as liberty of publication, tolerant church and temperate climate, a large majority were found in the triangular space in question, and there alone. The different nations vary at the different epochs in their scientific productiveness;[4] and he elaborately shows how closely the variation depends on some or other of the eighteen influences becoming favourable or unfavourable. The author, himself descended from the Huguenots, lays just stress on the influence of religious refugees, whose traditions were to work in a

[4] The authors tables of the scientific productiveness per million, of different nations at different times, are affected by a serious statistical error. He should have reckoned per of men above fifty, instead of the population generally. In a rapidly increasing country like England, the proportion of the youthful population to those of an age sufficient to enable them to become distinguished, is double what it is in France, where population is stationary; and injustice may be done by these tables to England in something like that proportion. They require entire reconstruction.

disinterested way for the public good, and at the same time to avoid politics. The refugees rarely had their property in land, of which the oversight occupies time, but in moveable securities thus they had leisure for work. Then, again, as they were debarred from local politics, the ambition, especially of those who had taken refuge in small countries, was to earn the approval of the enlightened men all over Europe, and this could most easily be effected by doing good work in science. Out of the ninety-two foreign associates of the French Academy, no less than ten were descended from religious refugees, usually in the third or fourth generation. Switzerland had eight out of the ten, and we may thence easily gather how enormously she is indebted to the infusion of immigrant blood. Similarly, the only two American associates —Franklin and Rumford—were descended from Puritans.

The blighting effect of dogmatism upon scientific investigation is shown both in Catholic and Protestant countries. The Catholics are the more dogmatic of the two, and they supply, in proportion to their population, less than one quarter as many of the foremost scientific men as the Protestants. There is not a single English or Irish Catholic among the ninety-two French foreign associates. Austria contributes no name, and the rest of Catholic Germany is almost barren. In Switzerland, the scientific productiveness of the Catholics is only 1-26th that of the Protestants. Again, the Catholic missionaries have done nothing for science, notwithstanding their splendid opportunities. In past days, when they were absolute masters of vast countries, as Paraguay and the Philippines, the smallest encouragement and instruction given at the college of the Propaganda to young and apt missionaries would have enriched Rome with collections of natural history. If any city more than others deserved to have the finest botanical garden and richest herbarium, it is Rome; but she has almost nothing to show.

The most notable instance of the repressive force of Protestant dogmatism is to be found in the history of the republic of Geneva. During nearly 200 years (1535 to 1725) its laity as well as clergy were absolutely subject to the principles of the early Reformers. Instruction was imposed on them; nearly every citizen was made to pass through the college, and many attended special courses at the Academy, yet during the whole of that period not a single Genevese distinguished himself in science. Then occurred the wane of the Calvinist authority, between 1720 and 1735. Social life and education became penetrated with liberal ideas;[5] and since 1739, the date of the first election of a Genevese to an important foreign scientific society—our own Royal Society—Geneva has never ceased to produce mathematicians, physicists, and naturalists, in a

[5] In 1786, public opinion had become so tolerant that it was enacted that candidates for the ministry should no longer be required to make a declaration of faith, but simply to promise to teach and preach conformably to the Bible and to the light of their own consciences (p. 204).

number wholly out of proportion to her small population.

The author argues from these and similar cases that it is not so much the character of the dogma taught that is blighting to science as the dogmatic habit in education. It is the evil custom of continually telling young people that it is improper to occupy their minds about such and such things, and to be curious, that makes them timid and indifferent. Curiosity about realities, not about fictions of the imagination, is the motive power of scientific discovery, and it must be backed up by a frank and fearless spirit. M. de Candolle, in spite of his anti-heredity declarations, enunciates an advanced pro-heredity opinion well worthy of note. He says it is known that birds originally tame, when found on a desolate island, soon acquire a fear of man, and transmit that fear as an instinctive habit to their descendants. Hence we might expect a population reared for many generations under a dogmatic creed to become congenitally indisposed to look truth in the face, and to be timid in intellectual inquiry.

Can, then, religion and science march in harmony? It is true that their methods are very different; the religious man is attached by his heart to his religion, and cannot endure to hear its truth discussed, and he fears scientific discoveries which might in some slight way discredit what he holds more important than all the rest. The scientific man seeks truth regardless of consequences; he balances probabilities, and inclines temporarily to that opinion which has most probabilities in its favour, ready to abandon it the moment the balance shifts, and the evidence in favour of a new hypothesis may prevail. These, indeed, are radical differences, but the two characters have one powerful element in common. Neither the religious nor the scientific man will consent to sacrifice his opinions to material gain, to political ends, nor to pleasure. Both agree in the love of intellectual pursuits, and in the practice of a simple, regular, and laborious life, and both work in a disinterested way for the public good. A strong evidence of this fundamental agreement is found in the number of sons of clergymen who have distinguished themselves as scientific investigators; it is so large that we must deplore the void in the ranks of science caused by the celibacy of the Catholic clergy. If Protestant ministers, like them, had never married, Berzelius, Euler, Linnaeus, and Wollaston would never have been born.[6] But to revert to what we were speaking about. There are some six different objects in the pursuit of which most men spend their energies; three of them refer to self—namely, property, pleasure, and political advancement; the other three imply devotion to ideas—namely, religion, science, and art. Without a doubt, as M. de Candolle says, the former three occupy one-half of the moral sphere of the human character, and the latter three the other.

It appears that the men distinguished in science have usu-

[6] Ed: Jöns Jacob Berzelius (1779–1848) and William Hyde Wollaston (1766–1828) were chemists.

ally been born in small towns, and educated by imperfect teachers, who made the boys think for themselves. Nothing is brought out more clearly in the work than that the first desideratum in scientific education is to stimulate curiosity and the observation of real things, and that too much encouragement of the receptive faculty is a serious error. The author justly laments that the art of observation is not only untaught, but is actually discouraged by modem education. Children are apt and eager to observe, but, instead of encouraging and regulating their instinct, the schoolmasters keep them occupied solely on internal ideas, such as grammar, the vocabularies of different languages, arithmetic, history, and poetry. They learn about the living world which surrounds them out of books, and not through their own eyes. One of the reformations he proposes is to make much more use of drawing as a means of careful observation, compelling the pupils to draw quickly the object they have to describe, from memory, after a short period allowed for its examination. He is a strong advocate for the encouragement of a class of scientific sinecurists like the non-working fellows of our colleges, who should have leisure to investigate, and not be pestered by the petty mechanical work of continual teaching and examining. Science has lost much by the suppression of the ecclesiastical sinecures at the time of the French Revolution, for there used to be many abbés on the lists of foreign scientific members, but they have now almost wholly disappeared. The modern ideas of democracy are adverse to places to which definite work is not attached, and from which definite results do not regularly flow. This principle is a wise one for the mass of mankind; but how utterly misplaced when applied to those who have the zeal for investigation, and who work best when left quite alone.

There is a curious chapter on the probability of English becoming the dominant language of the world in fifty or a hundred years, and being the one into which the more important scientific publications of all nations will, as a matter of course, be translated. It is not only that the English-speaking population will outnumber the German and the French, as these now outnumber the Dutch and the Swedish, but that the language has peculiar merits, through its relationship with both the Latin and the Teutonic tongues. It also seems that in families where German and French are originally spoken, French always drives out the German on account of its superior brevity. When people are in a hurry, and want to say something quickly, it is more easily said in French than in German. Precisely in the same way English beats French. Our sentences dont even require to be finished in order to be understood, because the leading ideas come out first; but as for old-fashioned tongues, their roundabout construction would be perfectly intolerable. Fancy languages like Latin and Greek, in which people did not say "yes" or "no." M. de Candolle is very disrespectful to classical Latin. He says that one must have gone through the schools not to be impressed by its ridiculous construction. Translate an ode of Horace literally to an unlettered artisan, keeping each word in its place, and it will produce the effect upon him of a building in which the hall-door was up in the third storey. It is no longer a possible language, even in poetry.

I have only space for one more of the many subjects touched upon in his book—that of acquired habits being transmitted hereditarily—and which has also formed the subject of a recent essay by Dr. Carpenter.[7] That some acquired habits in dogs are transmitted appears certain, but the number is very small, and we have no idea of the cause of their limitation. With man they are fewer still; indeed it is difficult to point out any one to the acceptance of which some objection may not be offered. Both M. de Candolle and Dr. Carpenter have spoken of the idiocy and other forms of nervous disorder which beyond all doubt afflict the children of drunkards. Here, then, appears an instance based on thousands of observations at lunatic asylums and elsewhere, in which an acquired habit of drunkenness, which ruins the will and nerves of the parent, appears to be transmitted hereditarily to the child. For my own part, I hesitate in drawing this conclusion, because there is a simpler reason. The fluids in an habitual drunkards body, and all the secretions, are tainted with alcohol; consequently the unborn

[7] Ed: Dr. William Benjamin Carpenter, 'On the Hereditary Transmission of Acquired Psychical Habits', *Contemporary Review* XXI (January 1873): 779–795.

child of such a woman must be an habitual drunkard also. The unfortunate infant takes its dram by diffusion, and is compulsorily intoxicated from its earliest existence. What wonder that its constitution is ruined, and that it is born with unstrung nerves, or idiotic or insane. And just the same influence might be expected to poison the reproductive elements of either sex. I am also informed, but have not yet such data as I could wish, that the children of recent teetotallers who were formerly drunkards are born healthy. If this be really the case, it seems to settle the question, and to show that we must not rely upon the above-mentioned facts as evidence of a once acquired habit being hereditarily transmitted.

FRANCIS GALTON

Chapter 4

Correspondence

4.1 1872/12/27. Francis Galton to Alphonse de Candolle

42, Rutland Gate,
London.
Dec. 27/72.

Dear Sir,

I thank you much for your volume which I received about a fortnight since and which I have read and re-read with care and with great instruction to myself.[1] Allow me to congratulate you on the happy idea of accepting the nominations of the French Academy and similar bodies as reliable diplomas of scientific eminence, and on thus obtaining a solid basis for your reasoning. I must however express no small surprise

[1] Candolle 1873.

at the contrast between your judgement on my theories and your own conclusions. You say and imply that my views on hereditary genius are wrong and that you are going to correct them; well, I read on, and find to my astonishment that so far from correcting them you re-enunciate them. I am perfectly unable to discover on what particulars, speaking broadly, your conclusions have invalidated mine. They have largely supplemented them, by thoroughly working out a branch of the inquiry into which I never professed to enter, but I literally cannot see that your conclusions, so far as heredity is concerned, differ in any marked way from mine. You say that race is all-important (p. 253 etc.)—that families of the same race differ from each other more widely than the races themselves (p. 268)—that physical form is certainly hereditary and that intellect is dependent on structure and must therefore be inherited (p. 326)—that for success, an individual must both "vouloir et pouvoir" (p. 92)[2]—that the natural faculties must be above mediocrity, (p. 106) and very many other similar remarks. I never said, nor thought, that special aptitudes were inherited so strongly as to be irresistible, which seems to be a dogma you are pleased to ascribe to me and then to repudiate. My whole book, including the genealogical tables, shows that ability—the "pouvoir"—may manifest itself in many ways. I feel the injustice you have done to me strongly, and one reason that I did not write

[2] 'be willing and able'.

earlier was that I might first hear the independent verdict of some scientific man who had read both books. This I have now done, having seen Mr Darwin whose opinion confirms mine in every particular. Let me, before proceeding to more agreeable subjects, complain of yet another misrepresentation. You say (p. 380) that I deny or doubt (contester) the good tendencies of children reared in the families of clergymen—I never said anything of the sort. What I did say was against the "pious," that is the over-religious. My genealogies are full of clergymen:—in the list you give, p. 381, I doubt if any of the parents are known to have been "pious"—though you might have quoted Haller in your favour.[3] Let me en passant remark about the last paragraph of your footnote, to p. 383, the sons of English clergy are [not] or were hardly ever sent into the army, because their parents could not afford it, and therefore their sons could not become Generals. Sometimes, but very rarely, they were put into the Navy, which is a less costly profession, and Admiral Lord Nelson was one of such.

I regret very much that you did not succeed in working out the genealogies of the scientific discoverers, on whom you rely, and on both sides. However there is no denying the fact that as a whole they are specialists, rather than illustrious men, and are therefore somewhat obscure to fame. Man against man, they would be nowhere in competition with a great statesman but they have owed more to concentration and the narrowing

[3] Albert von Haller. See Galton 1869, 213.

of their faculties than to a general prodigality of their nature. Such men are more easily affected by circumstances than the born geniuses about whom I chiefly busied myself, and are therefore all the more suitable subjects for an inquiry like yours, into the effects of different circumstances.

One of the most striking things to me in your book is the chilling influence on scientific curiosity you prove to result from religious authority. The figures you give seem to me of the highest importance. I am also greatly impressed with the conditions of fortune (funds not land) and the desire for an European rather than a local reputation which you ascribe to religious and other refugees. Switzerland's reputation seems made by the Huguenots, Euler and Haller being the only two in your list of purely native birth. I wish you had given the genealogies of the rest in full. Have you not made some slip of the pen in p. 125 at the bottom? If you cut off the sons of pasteurs I do not find that equality is re-established, nearly. Then see p. 40. where out of 20 fathers of associates only 4 were pasteurs and of all these associates only one was Catholic. There remain 14 non-clergy and 2 Quakers as parents of 16 Protestants to that 1 Catholic. Is not 'Protestant' a deceptive word? I fear most of the scientific men would be more truly described as 'infidel' or 'agnostic.'

How remarkable are your conclusions about teaching. I suppose severe teaching sacrifices many original minds but raises the level. We in England are in the throes of educational reform; wanting to know how best to teach "How to observe."

In your table XI of the scientific value of a million of different races, I note, what appears to me, a serious statistical error. You disregard the fact that some populations increase faster than others and have therefore always a plethora of children and of persons too young to be academicians. Take as sample and not very incorrect figures, that America U.S. doubles in 25 years, England in 50 and that France remains stationary. Then your calculation would do about a four-fold injustice to America, and a double injustice to England as compared to France, because it is at the age of 50 or thereabouts that people become academicians. The true comparison would be with the number of persons in the nations above the age of 50. This would avoid another great source of error arising from the very different chances of life of a child in different countries.

I fear the English physiologists will exclaim at your "état momentané lors de la conception."[4] Am I doing you an injustice in supposing that you argue on the hypothesis that conception and copulation are simultaneous? I certainly understand you do (pp. 311-2) but how can the argument stand? The spermatozoa do not get at the ovum for hours, perhaps many days, after copulation, and the ovum itself, when fecundated, has long been detached from the ovary.

[4] 'Les exceptions s'expliquent par la diversité des parents, leur état momentané lors de la conception'. 'The exceptions are attributable to the diversity of parents, their ephemeral state during conception'.

I feel, now that I have come to the end of this letter, that I have done little else than find fault, but I beg you to be assured that my general impression of the book is of another kind. I feel the great service you have done in writing it, and I shall do what I can to make it known, as it ought to be, in England. Can you get any facts out of Foundling Hospitals about heredity? The people here who administer ours are not scientific.

I have written an audacious article for Fraser's Magazine in Jan./73 of which I will send you a copy.[5]

Believe me, faithfully yours,
Francis Galton.

4.2 1873/01/02. Alphonse de Candolle to Francis Galton

Geneva,
2 January 1873.

Sir and esteemed colleague,

The volume I published has given you a combination of pleasant and unpleasant impressions. I can say as much of your letter of December 27th but before discussing certain points, I wish to make a general disclaimer. Unless it escapes me, in the 482 pages of my book a sentence or even a word

[5] Galton 1873a.

that may cast doubt on my respect for your impartiality, your character and your talent for inquiry, can absolutely only be in error and contrary to my intentions. You have always sought the truth. I enjoyed your work a lot and if it was accepted practice to include extensive material from one author I would have quoted you more often.

The idea of consulting the appointments by the Academies came to me 40 years ago! I had asked one of my friends to obtain from the Institute's secretariat the lists of Foreign Associates and Correspondents between 1750 and 1789. More recent names are easy to find elsewhere. In 1833 I wrote a memoir on the lists of Paris and the Royal Society. I did not publish it then because it was a little presumptuous for a young man to assess the value of illustrious scholars, among whom were his father and some distinguished men besides. Once I appeared on some lists myself, I was reluctant to discuss the matter. Finally, at the age of sixty-six, after a series of accomplishments that justified my position, courage came to me, and I dared to rise above personal considerations of any kind.

My writing was very far along when I discovered your work. I read it with infinite pleasure, and I have just re-read the most important chapters.

We are nicely in agreement about the facts. We have the same ideas about races. You have considered a larger number of categories of men. I studied the scholars in a more particular way, employing a different method, but with noticeably close results to yours.

I hold to the belief that there is not an opposition, just a noticeable difference in assessing the causes that have influenced the facts.

You usually highlight, as the main cause, heredity. When you speak of other causes they are mentioned in passing without trying to unravel their particular importance, individually or in combination. From time to time you do mention these other causes. Thus one can read many pages where you demonstrate the influence of heredity before meeting a line, as at the top of page 88, on social influences. The very title of the work implies the idea of looking only at heredity, its laws and its consequences, otherwise you would have said: On the effect of heredity and other circumstances as to genius. Surely you rendered a real service to science, but your point of view was essentially hereditarian.

For my part, I had more to follow up on. It has not been difficult for me to confirm the influence of heredity with further observations, but I have never lost sight of other causes, and the rest of my research has convinced me of general factors more important than heredity, at least among men of the same race. If we compare negroes with whites, or even Asian whites with European whites, the effect of race is predominant, but among the men of our civilized countries the effect of traditions, examples and influences within families seemed to me to exert more influence than heredity per se. After that comes general education, public opinion, institutions, etc. I have endeavored to distinguish the particular influence due to all these causes, which varies according to country and epoch, and which assists or limits the effects of heredity. The purpose of my research was therefore distinct from yours and the results were different without being contradictory. That's what I said at p. 93.

Now for some detailed observations.

At the bottom of p. 273 you speak of facts and some common opinion unfavorable to the children of religious people.[6] In my opinion the vast majority of clergymen are pious, and as I noticed, however, cases in which their children had gone wrong, I examined the facts, and found (to my surprise) that a large proportion of famous scientists had been sons of pastors or ministers. I drew an argument for the importance of fundamental moral education.

I did not understand your observation about a numerical error at p. 40. Counting again on the last column of pp. 36-40, I find 13 sons of pastors or ministers as I said. By parenthesis, I sometimes suspected that Sir David Brewster was a 14th. He was the son of a <u>Rector</u> of the Grammar School of Jedburgh, which in that country may well be a minister. A notice in a religious journal informed me that Sir David had been raised

[6] Galton 1869, 273.

in a very pious environment.[7]

I have not been able to determine the family origin of Mr Owen, just as I do not know that of Mr Airy, whom the Academy of Paris has nominated since 1869 as a foreign partner. Agassiz, also named since my collection, is the son of a Swiss pastor (of indigenous family).

If we subtract from the number of Swiss scientists those who come from foreign families, there would still remain a respectable number who would rank our little country next to the Scandinavian States and Holland, according to the period. It would not be fair in itself, because our scholars of foreign origin were all born in Switzerland, and even grandchildren or great-grandchildren of refugees born in Switzerland. For them the influence of heredity had already been greatly attenuated by the geometrical law of degrees.

Certainly the tree which Calvin and his friends had planted at Geneva, with its branches of Holland, Scotland, English and American puritans, was endowed with great vigor. Our strain in Geneva changed in a liberal sense in 1720 (pp. 127 and 205), as it did later in Boston and even a little in Holland and Scotland, but there remained in all these countries a spirit of independence and a stubbornness that was favorable to the sciences when individuals took them up. I did not want to dwell on such a small country as Geneva, but here are some facts

[7] Sir David Brewster (1781-1868) was a Scottish physicist. He was educated for the ministry.

that may interest you. At the time of the Reformation many noble families left Geneva to remain Catholics. There came instead a host of gentlemen and learned bourgeois from France and Italy who were zealous for the new religion. Thanks to their antecedents and their education they entered the class of the notable families of the country, which they dominated and they became the foundation of a local aristocracy which continued in fact, without title or legal sanction, until 1841. Not only in government and councils, but originally through the offices of pastors, and in the eighteenth century and up to 1841 through the professors of the Academy (with or without teaching duties) who formed reputable public education. Thanks to these practices a studious young man, of a notable family, could at least make a living. He also had a good chance to marry into money. You can see what encouragement existed for the sciences, specifically in the families of former refugees.

I could have given more information about the mothers and other forebears of Swiss scientists, especially from Geneva, but you have shown that past the immediate stages, heredity has very little influence. To tell the truth, all our somewhat old families in Geneva have Huguenot blood (if we dare to use this old phrase after your curious work on transfusion), only this blood is quite diluted. It is the institutions and mores implanted by our refugee ancestors that have persisted in influence much longer than heredity. Now an influx of Catholic workers from neighboring countries and a series of revolutions have given us a new social state. We become Americans. Nearly the only distinction between families will become wealth. That will not be for the benefit of science. Other cantons of Switzerland (Zurich, Vaud, Neuchâtel) are preparing to eclipse us, with improving local conditions.

For the proportions of scientists with respect to the populations of their countries, it would have been better to calculate the men of a certain age, either the one at which we are ordinarily elected at the Academies, or the one at which we begin to usefully work. Unfortunately it was impossible for the years 1750 and 1789, which I took care of, and very difficult for 1829 and even 1869. Sweden is the only country that had an age census in the last century. For the rest of Europe I was forced to make estimates even for the total population. We could not have the age distribution, in 1829 and 1869, in countries like Germany and Italy, where each state made its own censuses at different times and often did not disclose the details. Note, moreover, that the proportion of child deaths is all the greater as there are more children to be cared for, so that the number of adults per country is not as variable as one might think.[8]

There would be an important correction to apply; that of deducting foreign-born individuals from each country and adding nationals who have settled elsewhere. In this way,

[8] Here Candolle is arguing that differing child mortality rates even up the number of adults across countries despite their gross differences in population size.

in the United States, in 1869, some 5 1/2 millions would be removed, which would have to be distributed mainly over the British domains and Germany. The correction would be fair, because if there had been American foreigners who were Academy members I would have assigned them to their country of birth. That would not place the United States in the lower reaches of my tables. And how do we know how many British subjects are settled on the Continent or elsewhere in America? That of the Germans established in Russia, France, etc.? The subject luckily does not require so much precision. As I have said many times, population figures do not correlate with the exceptional groups of men engaged in science. It is necessary to include in each country considerable parts of the population which play an insignificant scientific role, such as Austria in Germany, the Kingdom of the Two Sicilies in Italy, Ireland in the United Kingdom, the Catholic Cantons in Switzerland. Population calculations can not, therefore, be of decisive statistical value, but they are useful in order to be able to appreciate the other causes which have operated in various countries, at different times, by taking into account the additional details necessary to qualify the statistics.

On fertilization, distinct from copulation, I do not see the value of the objections well. There is, in what I have said, 1. facts, 2. conjectures. The facts are that men in a state of intoxication (a temporary affection of the brain) have often produced idiots, epileptics, & c. and that a bitch heavily injured on the rear of the back during mating gave rise to small defectives in the spine (Lucas 2, p.250) Thus in man the spermatozoa can be modified by the morbid state of the nervous system, and in the canine species the ovules can be modified by an accident which occurred at the moment of the copulation. These are facts. Now it does not seem improbable to me to think that other temporary events could also influence alcoholism or an injury. In the human species a terror, a dominant or exclusive idea (monomania species) can last for several days and affect not only the spermatozoa but also the eggs when they become detached. I remember that at the time of the siege of Sevastopol there were people who had insomnia caused by the dreadful suffering reported by the newspapers. The events of 1870-71 caused a great deal of mental alienation and certainly disturbed many minds to a lesser degree. I would not be surprised if this was not a cause of increase of madness or idiocy in children born in 1871 in a part of Europe. In countries where wealth must be shared equally among children, you can not imagine the more or less secret terror of many women at the thought of a new pregnancy. Do you also consider the nervous state of some women when they were unfaithful to their husbands and that of a husband who hates his wife without revealing it, as in the case of the Duke of Praslin for example, who maintained his marital relations with the intention of murdering the Duchess.[9]

[9] Charles Laure Hugues Théobald, duc de Choiseul-Praslin (1805–1847)

I do not know of any study here about the foundlings or how to go about one. These children come from unknown and very varied parents. I do not know what one could conclude from their abilities. Rousseau put his children in the foundling hospital. We wondered if they had become something. I suspect they died young, considering the detestable conditions of the old hospitals of Paris.

I will be very obliged to be sent the article in Fraser's Magazine of which you speak, as well as of any writing with or without criticism of my work which you would have the goodness to publish. In the meantime, I beg you to believe me, sir, your very devoted colleague.

Alph. de Candolle.

P. S. I had the publisher send a copy of my book to the Royal Society. I hope it has arrived.

Genéve,
2 Janvier 1873.

Monsieur et honoré collègue,

Le volume que j'ai publié vous a causé un mélange d'impress-ions agréables et désagréables. Je puis en dire autant de votre lettre du 27 Décembre mais avant de discuter certain points, je desiré vous faire une déclaration générale. S'il m'échappe, dans les 482 pages de mon livre, une phrase, un mot pouvant

stabbed his quarrelsome wife to death in 1847 then committed suicide when in custody by overdosing on laudanum or some other poison.

faire douter de mon respect pour votre impartialité, votre caractère et votre talent d'investigation, ce ne peut étre absolument que par erreur et contrairement à mes intentions. Vous avez toujours cherché la vérité. J'ai apprécié beaucoup votre travail et s'il n'était pas inusité de transcrire de nombreux articles d'un auteur je vous aurais cité encore plus souvent.

L'idée de consulter les nominations par les Académies m'est venue il y a 40 ans! J'avais prié un de mes amis de prendre au sécretariat de l'Institut les listes des Associés étrangers et Correspondants de 1750 en 1789. Les noms modernes sont aisés à trouver ailleurs. J'avais redigé en 1833 un mémoire sur ces listes de Paris et sur celles de la Société Royale. Si je ne l'ai pas publié alors c'est qu'il me semblait un peu présomptueux chez un jeune homme de mesurer ainsi la valour de savants illustres, parmi lesquels se trouvait son père et quelques hommes distingués a côté de lui. Une fois moi-même sur certaines listes, it me répugnait d'en parler. Enfin, à 66 ans, après une série do travaux spéciaux propres à justifier ma position, le courage m'est venu et jai penaé pouvoir m'élever au dessus des considérations personnelles de toute nature.

Ma rédaction était fort avancée quand j'ai connu votre ouvrage. Je l'ai lu avec infiniment de plaisir, comme je viens d'en relire les chapitres les plus importants.

Nous sommes admirablement d'accord sur les faits. Nous avons les mêmes idées sur les races. Vous avez envisagé un plus grand nombre de catégories d'hommes, mais celle des savants que j'ai étudiee d'une manière plus spéciale, avec une méthode différente, m'a donné de résultats extrêmement semblables aux vôtres quant aux faits.

Je persiste à croire qu'il y a, non pas une opposition mais une difference assez sensible dans l'appreciation des causes qui ont influé sur les faits.

Vous faites babituellement ressortir, comme cause princi-pals,l'héredité. Quand vous parlez des autres causes elles sont indiquées accessoirement et sans chercher à démêler ce qui tient particulièrement à elles ou à chacune d'entre elles. De loin en loin vous mentionnez ces autres causes. Ainsi on peut lire bien des pages où vous démontrez l'influence de 1'hérédité avant de rencontrer une ligne comme au haut de la page 88 sur les social influences. Le titre même de l'ouvrage implique l'idée de rechercher uniquement sur l'hérédité, ses lois et ses conséquences, autrement vous auriez dit: On the effect of heredity and other circumstances as to genius. Assurément vous avez rendu un vrai service à la science, mais votre point de vue était essentiellement l'hérédité.

Quant à moi j'ai eu davantage de venir après vous. Il ne m'a pas été difficile de confirmer par de nouveaux faits l'influence de l'hérédité, mais je n'ai jamais perdu de vue les autres causes, et la suite de mes recherches m'a convaincu qu'elles ont en général plus d'importance que l'hérédité, du moins parmi les hommes de même race. Si l'on compare des negres avec des blancs, on même des blancs asiatiques avec des blancs européens, l'effet de la race est prédominant, mais parmi les hommes de nos pays civilisés l'effet des traditions, exemples et conseils dans l'intérieur des familles m'a paru exercer plus d'influence que l'hérédité proprement dite. Vient ensuite l'education extérieure, l'opinion publique, les institutions etc. Je me suis applique à distinguer la part d'influence de toutes ces causes, part qui varie suivant les pays et les époques, et qui favorise ou contrarie lee effets de l'hérédité. Le but de mes recherches était donc différent du vôtre et les résultats en ont été diférents sans être opposés. C'est ce que j'ai dit à la p. 93.

J'en viens aux observations de détail

Au bas de la p. 273 vous parlez de faits et d'une certaine opinion courante défavorables aux enfants de personnes religieuses (religious). Dans mon idée l'immense majorité des clergymen est religieuse, et comme j'ai remarqué cependant moi-même des cas dans lesquels leurs enfants avaient mal tourné, j'ai examiné les faits, et j'ai trouvé (à ma grande surprise) qu'une forte proportion de savants célèbres avaient été fils de pasteurs ou ministres. J'en ai tiré un argument en faveur de l'importance d'éducation simple et morale.

Je n'ai pas compris votre observation sur une erreur de chiffre à la p. 40. En comptant de nouveau sur Ia dernière colonne des pp. 36-40, je trouve 13 fils de pasteurs ou ministres comme je l'avais dit. Par parenthèse, j'ai soupconné quelquefois que Sir David Brewster était un 14ème. Il était fils d'un Rector of the Grammar school of Jedburgh, ce qui d'après le pays peut faire croire qu'il était peut-être ministre. Une notice dans un journal religieux ma appris que Sir David avait été élevé dans une atmosphère très pieuse.

Je n'ai pas pu savoir l'origine de famille de Mr Owen, de même que j'ignore celle de Mr Airy, que l'Académie de Paris a nommé depuis 1869, associé étranger. Agassiz, nommé également depuis mon tableau, est fils d'un pasteur suisse (de famille indigène).

Si lon retranchait du nombre des savants suisses ceux qui descendent de familles étrangères, il resterait encore un nombre assez respectable qui placerait notre petit pays à côté des Etats Scandinaves et de la Hollande, selon les époques. Ce ne serait pas juste en soi, parce que nos savants dorigine étrangère étaient tous nés en Suisse, et même petit-fils ou arrière-petit-fils de réfugiés nés en Suisse. Pour eux linfluence de lbérédité avait déjà été atténuée énormément par la loi géométrique des degrés.

Assurément larbre que Calvin et ses amis avaient planté à Genève, avec ses rameaux de Hollande, dEcosse, des puritains anglais et dAmérique, était doué dune grande vigueur. Notre souche à Genève sest modifiée dans un sens libéral en 1720 (p. 127 et 205), comme plus tard à Boston et même un peu en Hollande et en Ecosse, mais il est resté dans tous ces pays un esprit dindépendance et une persistance de volonté qui ont été favorables aux sciences lorsquil a convenu aux individus de sen occuper. Je nai pas voulu métendre davantage sur un aussi petit pays que Genève, mais voici quelques faits qui peuvent vous intéresser. A lépoque de la Réformation beaucoup de familles nobles quittèrent Genève pour demeurer catholiques. Il vint à la place une foule de gentilhommes et bourgeois instruits de France et dItalie, qui étaient zélés pour la nouvelle religion. Grâce à leurs antécédents et à leur éducation ils entrèrent dans la classe des familles notables du pays, quils dominèrent et ils devinrent le fond dune aristocratie locale qui a subsisté de fait, sans titre ni privilège légal, jusquen 1841. Cette sorte de patriciat ne visait pas seulement aux places du gouvernement et des Conseils ; elle occupait à lorigine les charges de pasteurs et dans le XVIIIe et jusquen 1841 celles de professeurs de lAcadémie (avec ou sans enseignement) qui donnait la surveillance de linstruction publique et un rang honorable dans lopinion. Grâce à ces murs un jeune homme studieux, dune famille notable, pouvait se contenter dune fortune médiocre. Il avait dailleurs une bonne chance pour se marier richement. Vous voyez quels encouragements existaient en faveur des sciences, précisément dans les familles danciens réfugiés.

Jaurais pu donner plus dinformations sur les mères et autres ascendants des savants suisses, surtout genevois, mais vous avez démontré que passé les premiers degrés lhérédité influe fort peu. A vrai dire toutes nos familles un peu anciennes à Genève ont du sang de huguenot (si lon ose employer cette vieille locution après votre curieux travail sur la transfusion), seulement ce sang est tout à fait dilué. Ce sont les institutions et les murs implantées par nos ancêtres réfugiés qui ont pesé bien plus longtemps que lhérédité. Mainte-nent une affluence douvriers catholiques des pays voisins et une série de révolutions nous ont donné un nouvel état social. Nous devenons américains. Désormais la distinction presque unique entre les familles sera la fortune. Ce ne sera pas au profit de la science. Dautres cantons de la Suisse (Zurich, Vaud, Neuchâtel) se préparent à nous succéder, les bonnes conditions étant mieux réunies chez eux.

Dans les proportions de savants à légard des populations de pays, il aurait mieux valu pouvoir calculer sur les hommes dun certain âge, soit celui auquel on est ordinairement élu aux Académies, soit celui auquel on commence à travailler utilement. Malheureusement cétait impossible pour les années 1750 et 1789, dont je me suis occupé, et fort difficile pour 1829 et même 1869. La Suède est le seul pays qui ait eu un dénombrement par âges dans le siècle dernier. Pour le reste de lEurope jai été forcé de recourir à des estimations même pour la population totale. On ne pourrait pas avoir la division par âges, en 1829 et 1869, dans les pays comme lAllemagne et lItalie où chaque Etat faisait ses recensements sur la base quil imaginait, à des époques différentes et souvent nen faisait pas ou nepubliait pas les détails. Remarquez dailleurs que la proportion des décès denfants est dautant plus forte quil y a plus denfants à soigner, doù il résulte un nombre dadultes moins différent quon ne croirait dun pays à lautre.

Il y aurait une correction importante à faire ; celle de défalquer de chaque pays les individus nés à létranger et dajouter les nationaux qui se sont établis ailleurs. On ôterait de cette manière aux Etats-Unis, en 1869, environ 514 millions, quil faudrait répartir sur les Royaumes britanniques et lAllemagne principalement. La correction serait équitable, car sil y avait eu dans les étrangers établis en Amérique des titulaires dAcadémie je les aurais imputés à leur pays de naissance. Cela ne sortirait pourtant pas les Etats Unis de la région inférieure de mes tableaux. Et comment savoir la quantité de sujets britanniques établis sur le Continent ou ailleurs quen Amérique ? Celle des Allemands établis en Russie, en France etc. ? Le sujet heureusement nexige pas une si grande précision. Je lai dit plusieurs fois, les chiffres de population ne sont pas en corrélation avec les groupes exceptionnels dhommes soccupant de science. On est forcé denglober dans chaque pays des parties considérables de population qui jouent un rôle scientifique insignifiant, comme lAutriche en Allemagne, le royaume des Deux Siciles en Italie, lIrlande dans le Royaume uni, les cantons catholiques en Suisse. Les calculs sur les populations ne peuvent donc pas avoir une véritable valeur statistique, mais il sont utiles pour pouvoir apprécier les causes qui ont influé en divers pays, à diverses époques, en tenant compte des détails accessoires propres à modifier limpression déterminée par les chiffres.

Sur la fécondation, distincte de la copulation, je ne vois pas bien la valeur des objections. Il y a dans ce que jai dit 1. des faits, 2. des conjectures. Les faits sont que des hommes en état divresse (affection temporaire du cerveau) ont engendré souvent des idiots, épileptiques, etc. et quune chienne fortement blessée sur larrière partie du dos pendant laccouplement a donné naissance à des petits défectueux du train de derrière (Lucas 2, p. 250)1. Donc chez lhomme les spermatozoaires peuvent être modifiés par létat momentané maladif du système nerveux, et chez lespèce canine les ovules peuvent être modifiés par un accident survenu au moment même de la copulation. Voilà des faits. Maintenant il ne me paraît pas hasardé de croire que dautres affections temporaires pourraient aussi influer comme lalcoolisme ou comme une lésion. Dans lespèce humaine une terreur, une idée dominante ou exclusive (espèce de monomanie) peuvent durer plusieurs jours et influeraient non seulement sur les spermatozoaires mais aussi sur les ovules au moment où ils vont se détacher. Je me souviens quà lépoque du siège de Sébastopol il y avait des personnes qui avaient perdu le sommeil de leffroi et de la commisération des souffrances racontées par les journaux. Les événements de 187071 ont causé beaucoup daliénations mentales et sûrement ont troublé à un moindre degré beaucoup desprits. Je ne serais pas étonné que ce nait été une cause

at ease, of a settled life, in the prime of life, whose wife and ancestors were sane, who had had a very healthy spirited girl, and who, having accidentally drunk a little too much wine, and having then approached his wife, had an idiotic son. He has told the story since and I know the people in question. Alcoholism is not the only temporary condition possible. This, then, is a cause of variations from one generation to the next, independent of ordinary heredity and atavism, that is to say, from that set of permanent former causes of which you speak with reason.

Your desire to foster beneficial family alliances and reduce the frequency of others is very natural, but you may not know well enough how special interests and the fear of harm will prevent the truth from coming to light. If there is any circumstance in which honest people distort the truth, it is that of the marriages concluded close to them. By sympathy, antipathy or desire to favor other individuals, they sometimes lie shamelessly. I know of only two ways of getting rid of unwelcome unions, one of extending the legal prohibitions against marriage based on of kinship, the other of proliferating public warnings about the dangers of consanguineous unions. You warn in that way, and in so doing perform a public service.

Receive, I beg you, my dear sir, the assurance of my very devoted affection.

Alph. de Candolle

Genève 29 avril 1873

Mon cher Monsieur

Jai des remerciements à vous faire pour votre article du Fortnightly review qui attire lattention du public sur mon dernier ouvrage. Si jaimais la polémique je répondrais à quelques unes de vos observations, mais cest un genre décrits pour le quel je nai que de la répugnance, et je préfère marrêter aux points nombreux sur lesquels nous sommes daccord. Probablement il viendra un moment où le libraire me demandera une seconde édition, alors je reverrai les points que vous signalez et modifierai plus ou moins, selon ce qui me paraitra convenable en tenant compte des faits.

Vos réflexions sur les tristes conséquences dun état fréquent divrognerie sur les facultés intellectuelles des enfants sont très justes. Quand il y a habitude divresse, cest-à-dire ivrognerie, la santé générale est atteinte, par conséquence celle des enfants qui naissent, mais ce qui est bien moins connu et cependant très curieux, un état exceptionnel, temporaire, divresse de lun des parents, peut avoir des conséquences pour lenfant procréé alors. Jen ai cité un exemple à Mr Darwin dans une lettre dil y a 2 ou 3 mois, mais je ne me rappelle pas si je vous en ai parlé. Cest le cas dun Monsieur, assez à son aise, dune vie réglée, dans la force de lâge, dont la femme et les ascendants étaient sains desprit, qui avait eu une fille très saine de corps et desprit, et qui ayant par hasard bu un peu trop de vin et sétant alors rapproché de sa femme a eu de ce fait un fils idiot. Il la raconté depuis et je connais les individus en question. Lalcoolisme nest pas la seule affection temporaire possible. Voilà donc une cause des variations dune génération à lautre, indépendante de lhérédité ordinaire et de latavisme, cest-à-dire de cet ensemble de causes antérieures permanentes dont vous parlez avec raison.

Votre désir de favoriser par de bons renseignements les alliances de familles bien portantes et de diminuer la fréquence des autres est très naturel, mais vous ne savez peut-être pas assez combien les intérêts particuliers et la crainte de nuire empêchera la vérité de se faire jour. Sil y a quelque circonstance où dhonnêtes gens faussent la vérité cest celle des mariages projetés autour deux. Par sympathie, antipathie ou désir de favoriser dautres individus, ils mentent quelquefois effrontément. Je ne connais que deux manières déloigner les unions fâcheuses, lune détendre les défenses légales de mariage pour cause de parenté, lautre de multiplier les avis au public sur les dangers dunions consanguines Vous agissez dans ce sens et rendez service autour de vous et même ailleurs.

Recevez, je vous prie, mon cher Monsieur, lassurance de mes sentiments très dévoués.

Alph. de Candolle

4.4 1873/05/07. Francis Galton to Alphonse de Candolle

42, Rutland Gate, London.
May 7/73

My Dear Sir,

It gave me much pleasure to receive your letter. I assure you I feel like yourself, that the subjects on which we differ are altogether subordinate to the common interest we have in arriving at the truth on the same line of inquiry. My article in the *Fortnightly* was much shorter than I should have liked to have made it, but there was a difficulty about space and I crammed all I could in what was given to my disposal.[11]

Of the many topics in your work left unnoticed I regretted much not being able to speak of your most just criticism of the misuse of the word 'Nature.' For my part, I will never offend again unless through a slip of the pen. Your work has been read by many of my scientific friends here, and a passage in it prompted one of the most effective parts of by far the most effective speech,—that of Dr Lyon Playfair,—in the recent Parliamentary debate upon Irish University Education. The debate, as perhaps you may have seen, was one of extreme importance to the future of science in Ireland, and the question was how far it should be submitted to or emancipated from

[11] See Galton 1873b, reproduced in Chapter 3, p. 61.

Figure 4.1: Alphonse de Candolle.

Figure 4.2: Augustin Pyramus de Candolle (1778–1841), father of Alphonse and founder of the botany dynasty.

Figure 4.3: Casimir de Candolle (1836–1918), son of Alphonse and also a distinguished botanist.

Catholic control. Lyon Playfair quoted the effect of Calvinism in Geneva on science, during the time of its ascendancy in wholly suppressing it, which was shown by the immediate start made by science as soon as the strict dogmatic influence began to wane. He spoke with excellent effect and success, and I know that he derived at least that part of his argument from you, because I had myself directed his attention to your work previously as having a direct bearing on his then proposed speech.

Thank you for your interesting fact about impregnation under the effect of alcoholism. One of course needs many such facts and it occurs to me that perhaps some direct experiment might be made, say with white mice, which breed very frequently and largely, are easily reared and cheap to keep. The he-mouse might be fed on some suitable narcotic stimulant before being put in with the female. I have no idea what stimulant would be suitable, one would have to try cannabis sativa, belladonna, opium, etc. A strong instance (if accurately recorded) of alcoholism combined both with the evil influences of close interbreeding and of old age on the part of one of the parents, in producing no bad effect on the offspring, is that of Lot and his daughters (Genesis xix. 31).

You are good enough to remark on my views about improving the human breed, showing the difficulty of detecting and of discovering defects which families scrupulously conceal. But then on the other hand, it must be borne in mind that my primary object is not to deter the bad from, but to encourage the good breeds in, making early marriages. Those who are conscious of being of a good stock would court inquiry, for by having a warranty they would be advantaged. People take such extraordinary pains to found families that they could easily be taught the importance of marrying their sons and daughters to persons likely to cooperate in begetting children capable of supporting the dignity of the family. Hence youths having warranties would be sought after far more than the same persons are sought after now. After many generations, the absence of a warranty would look suspicious. Encouragement of the best is the surest and safest way of discouraging the inferior. We are such a set of mongrels that except in extreme cases we should not be justified in 'banning' any marriage. All we can say is, that some marriages are more hopeful than others. I therefore go no further at present, than urging that hopeful marriage should be encouraged.

If an autumn's tour should take me to Geneva, I trust you will not think it a liberty if I do myself the pleasure of seeking your personal acquaintance, with a view to some conversation on the many subjects in which we have a strong common interest.

Believe me very faithfully yours,
Francis Galton.

4.5 1873/06/16. Alphonse de Candolle to Francis Galton

16 June 1873.

My dear sir,

You make me hope for a visit in the course of the autumn and as I would be very sorry to miss you, let me tell you that I will probably be in Geneva from the middle or at the latest by the end of September. I intend to be away for the month of August, but it is doubtful whether my excursion will be longer than 5 or 6 weeks.

If you pass through our city you have only to ask for me at Court St Pierre 3 (on the 2d floor). This is my herbarium and also my library. I come here every day from 11 am to 2 pm, except Sundays.

In the evening and on Sundays you would find me in Vallon near Chêne 2 miles from the city. A tram exists between Geneva and Chêne and from Chêne to my country house it is 8 minutes.

Your idea of experimenting on mice would be excellent to put into practice. It should first be seen what poisonous substance these animals are willing to eat, and how much they can be given without hurting them too much, yet producing sensible effects. Cannabis has less inebriating matter in the seed than in the leaves, but there must be some. An alcoholic mash may be more convenient, because one would know the dose of alcohol administered.

Pigeons that mate under the influence of alcohol might yield small tumblers? Mr. Darwin could try it better than anyone. But he might say that the fertilization of eggs in birds is surrounded by too much uncertainty. It may be that not exactly those eggs were fertilized at that time by the alcoholic male.

Receive, my dear sir, the assurance of my eager greetings.

Alph. de Candolle.

16 Juin 1873.

Mon cher Monsieur,

Vous me faites esperer une visite dans le courant de lautomne et comme je serais bien fache de vous manquer, permettez-moi de vous dire que je serai sans doute a Geneve depuis le milieu ou au plus tard la fin de septembre. Jai 1intention de mabsenter pendant le mois daout, mais il est bien douteux que mon excursion se prolonge au dela de 5 ou 6 semaines.

Si vous passez dans notre ville vous navez qua me demander Cour St Pierre 3 (au 2d etage). Cest la quest mon herbier et aussi ma bibliotheque. Jy viens tous les jours de 11 a 2 h., excepte le dimanche.

Dans la soiree et les dimanches vous ne trouveriez au Vallon pres Chene à 2 milles anglais de la ville. Un tramway existe entre Genève et Chêne, et de Chene a ma maison de eampagne il y a 8 minutes.

Votre idee d'Experimenter sur les souris serait excellente a suivre. Il faudrait voir dabord quelle substance toxique ces animaux mangent volontiers et quelle dose on peut leur en donner sans leur faire trop de mal et en produisant cependant des effets sensibles. Le Cannabis a moins de matière inébriante dans la graine que dans les feuilles, mais il doit pourtant y en avoir. Une pâtée alcoolisée serait peut-être plus commode, parce quon saurait bien la dose dalcool administrée.

Des pigeons qui saccoupleraient sous lempire de lalcool donneraient peut-être des petits tumblers ? Mr Darwin pourrait lessayer mieux que personne. Mais il dirait peut-être que la fécondation des ufs dans les oiseaux est entourée de trop dincertitudes. On ne saurait peut-être pas exactement quels ufs auraient été fécondés à tel moment par le mâle alcoolisé.

Recevez, mon cher Monsieur, lassurance de mes salutations empressées.

Alph. de Candolle.

4.6 1873/09/30. Alphonse de Candolle to Francis Galton

Geneva Sep 30, 1873

My dear sir

As I had written to tell you that I would be at home in autumn, except perhaps for a short absence in Paris, I have to warn you that, indeed, I will go there on October 13 or 14 and will stay a week or so at the Hotel du Louvre. If by chance you go to Paris at this moment we could see each other there. After that I will return to Geneva where you will find me in my library every day, as usual, if you go to Switzerland.

In the hope of meeting you sooner or later, I am always, my dear sir, your very

Alph. of Candolle

Genève 30 sept. 1873

Mon cher Monsieur

Comme je vous avais écrit pour vous dire que je serais chez moi en automne, excepté peut-être une courte absence à Paris, je viens vous prévenir que, effectivement, jirai à le 13 ou le 14 octobre et y resterai une huitaine de jours, à lhôtel du Louvre. Si par hasard vous passez à Paris à ce moment nous pourrions nous y voir. Après je reviendrai à Genève où vous me trouveriez dans ma bibliothèque chaque jour, selon mon habitude, si vous vous dirigez vers la Suisse.

Dans lespérance de vous rencontrer tôt ou tard, je suis toujours, mon cher Monsieur, votre très évoué

Alph. de Candolle

Figure 4.4: Francis Galton in his late 30s, taken sometime in the mid to late 1850s.

Figure 4.5: The American popular edition of Galton's English Men of Science *(1874), written as a riposte to Candolle.*

TRANSACTIONS

OF THE

AMERICAN PHILOSOPHICAL SOCIETY

HELD AT PHILADELPHIA
FOR PROMOTING USEFUL KNOWLEDGE

NEW SERIES—VOLUME 65, PART 5
1975

A GUIDE TO FRANCIS GALTON'S
ENGLISH MEN OF SCIENCE

VICTOR L. HILTS
Associate Professor of the History of Science, University of Wisconsin

THE AMERICAN PHILOSOPHICAL SOCIETY
INDEPENDENCE SQUARE
PHILADELPHIA

August, 1975

Figure 4.6: In the early 1970s Victor Hilts discovered the key to English Men of Science *in a desk drawer in the Galton laboratory at UCL.*

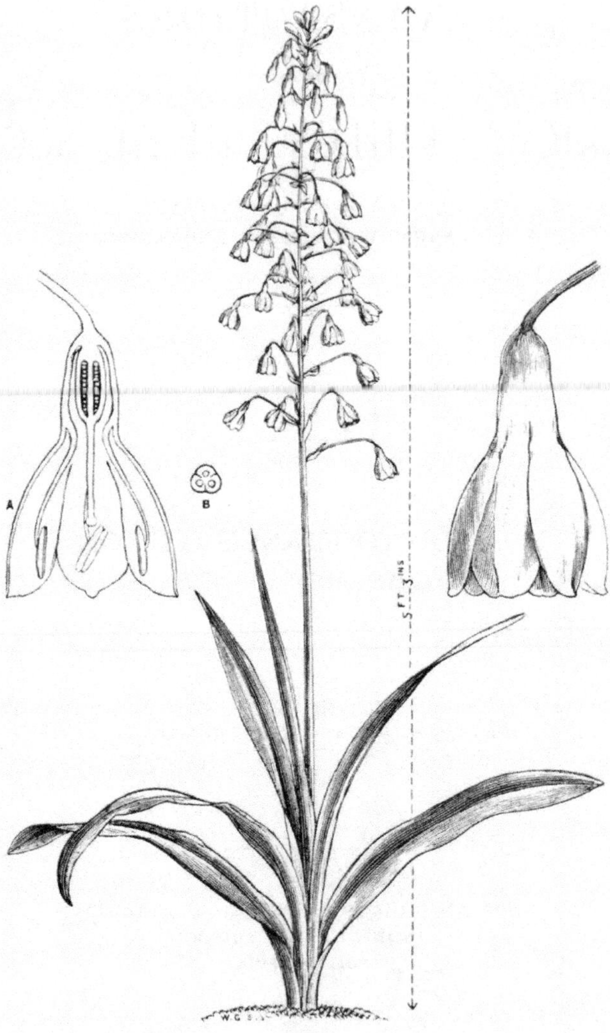

Figure 4.7: Galton's only connection to Candolle's field of botany is through the 'summer hyacinth', Galtonia Candicans—*named, to his surprise, in his honour.*

4.7 1873/10/18. Francis Galton to Alphonse de Candolle

42, Rutland Gate, London.
Oct. 18/73

My Dear Sir,

Thank you very much for your kind letter this morning telling me of your whereabouts this autumn. I heartily wish I could have managed to meet you in Paris or elsewhere, for I have much in hand at this moment concerning the topic which interests both of us so much, and which I hope to publish in 3 or 4 months.

My wife and myself passed our summer in the heart of Germany, in the Thuringer Wald, and I there continually consoled myself with your prophecy that that tiresome German language is doomed to extinction, as one of the dominant tongues. That, and our atrocious English spelling! for which I, not being a classical scholar, entertain no respect whatever!

Believe me very faithfully yours,
Francis Galton.

I have just left the house of a friend, where I had paid a short visit at the same time as our mutual friend Dr Hooker, who very shortly will occupy the most distinguished of English scientific posts, namely the Presidency of the Royal Society. He will be a most acceptable President over us.

4.8 1874/05/05. Francis Galton to Alphonse de Candolle

May 5/74

My dear Sir,

A few weeks back I gave a lecture at our Royal Institution on the subject that interests both of us, and only delayed sending you a printed account of it in the hopes of sending at the same time another and different memoir, which however will not be in print for some time.[12]

My lecture will reach you by book post, and I hasten to send it because I see that the Revue Scientifique has been so good as to publish a translation in French which however does not render some phrases quite exactly and gives a small but decided modification to their meaning. When the book will be complete, to which this lecture is a prelude, I cannot say; but as soon as it is out, I will send you a copy, for acceptance. It seemed to me well worth while to select, as I have done, a group of men of the same nationality, similar education, race, religion and period, in order to eliminate the disturbing influences of as many large variable causes as possible, and to bring out into stronger relief the effect of the residuum. I think you will be interested, when all the results are before you, in tracing the differences between them and your conclusions

[12] Galton 1874b.

derived from the study of a selection of much more able men but under more varied circumstances. My scientific men, for example, are mostly born in towns, but of every 5 of these men, 1 is born in London, 1 in other very large towns, 1 in moderately sized towns and 2 in villages or in country houses. The heredity comes out very markedly and follows unexpected and peculiar "laws" (if I may be allowed, for want of a better, to use so grand an expression). I have not touched on these in the lecture, there was no time.

The other memoir alluded to, at the opening of the latter, is a partially successful attempt to solve that very difficult mathematical question to which you drew attention in your book and about which I had plagued my not very brilliant mathematical head, at intervals, for many years namely the extinction of families by the ordinary laws of chance. I contrived to state the problem, in a not unreasonable form that was at the same time fitted for mathematical investigation and did my best to persuade friends to work it out. At length one friend has got the thing into a shape that admits of some general conclusions being drawn but it is by no means a "solution" of the problem in the ordinary sense of that word. When it is printed, I will send you copies. Perhaps you may persuade Swiss mathematicians to investigate further. It really ought to be solved if possible, but it is only too probable that a direct solution is an impossibility.

Our mutual friend Mr Bentham is, I am glad to say, abroad

on a holiday with the Hookers. His domestic griefs due to the long continued mental ill-health of his wife seem to have preyed much upon his spirits.

Fearing that I have wearied you by this long and somewhat egostistic letter I will now conclude and beg you to believe me faithfully yours,

Francis Galton.

4.9 1874/09/11. Alphonse de Candolle to Francis Galton

11 Sept. 1874.

My dear sir,

Several absences and indisposition of a few weeks prevented me from writing to you about your session at the Royal Institution: On men of science etc. Yet I read your booklet twice, with much pleasure. The foundation on which you rely is original. I doubt whether we could obtain so many answers elsewhere as in England, or as frankly and conscientiously. Starting from the list of the Royal Society I fear somewhat that the number of Scots was not high enough. As their education was formerly very different from that of the English that would be regrettable. No doubt you will speak of this difference in your book, which I eagerly anticipate reading.

Many of the conditions that your correspondents thought favorable would be found outside the Anglo-Saxon countries, but to a greater or lesser degree, and some would probably not be found at all.

Energy—It must be general, only perseverance sometimes predominates.

Health—I have known many scientists in poor health. When it affects the limbs, but nevertheless the head is good, bad health does not necessarily prevent success. When I was studying law my best teacher, a very skilful lawyer, was crippled in the legs and arms. At this moment, in Geneva, one of the most skilful men, a horticulturist and naturalist, is a former pastor. Life has gradually crippled his body from his extremities to his head for 25 years. He can not convey his food to his mouth, but he continues to read scientific books in 3 languages and dictates translations. He transports his torso (the only living part) to his garden to check the sequence of experiments he has ordered. The first of these two remarkable cripples had a big head and resembled Napoleon I; the second, also energetic, has a small one. I have often seen bodily exercises divert studies or destroy the habit of carefully observing. No doubt they must be proportionate to the degree of strength of each, but not beyond.

Practical business—This is typically English! On the Continent you would find a crowd of scientists who do not understand anything about the matter, who make fun of it, and whose fortune diminishes rather than increases. Neglect of business is more common among men of letters. I believe that a scientist rarely succeeds if he does not order his papers, his work and even in the ordinary habits of life, but that is not what you call practical business (commercial aptitude, industrial, etc.).

Your Reflections p. 6 ... All tends to ... divine, are extremely accurate, especially the anti-feminine character![13] I will not say why, it would offend the most beautiful half of our species.

I second the advantage of a varied education. Contrary to the opinion of college pedants, I believe with you that it is a great source of curiosity and other strengths, which are then developed in the specialism of a career. Only one must not continue his varied education indefinitely. You have to know how to become specialized.

In the hope of soon reading the projected volume of your interesting researches, I am always, my dear sir, your very devoted colleague

[13] 'All tends to show that the scientific mind is directed to facts and abstract theories, and not to persons or human interests. The man of science is deficient in the purely emotional element, and in the desire to influence the beliefs of others. Thus I find that two out of every ten do not care for politics at all; they are devoid of partisanship. They school a naturally equable and independent mind to a still more complete subordination to their judgment. In many respects their character is strongly anti-feminine. It is a curious proof of this, that in the very numerous answers which have reference to parental influence, that of the father is quoted three times as often as that of the mother. It would not have been the case, judging from inquiries I elsewhere made, if I had been discussing literary men, commanders, or statesmen, or, still more, divines', Galton 1874b, 232.

Alph. of Candolle.

11 Sept. 1874.

Mon cher Monsieur,

Plusieurs absences et une indisposition de quelques semaines mont empêché de vous écrire au sujet de votre séance de la Royal Institution: On men of science etc. Jai pourtant lu deux fois votre opuscule, avec beaucoup de plaisir, La base sur laquelle vous vous appuyez est originale. Je doute quon pût obtenir ailleurs quen Angleterre un aussi grand nombre de réponses faites franchement et consciencieusement. En partant de la liste de la Société royale je crains un peu que le nombre des Ecossais nait pas été assez élevé. Comme leur éducation était naguère très différente de celle des Anglais ce serait regrettable. Sans doute vous parlerez de cette différence dans votre ouvrage que je me réjouis beaucoup de lire.

Plusieurs des conditions réputées favorables par vos correspondants se retrouveraient certainement hors des pays anglosaxons, mais à des degrés autres et quelques-unes ne se retrouveraient probablement pas du tout.

Energie - Ce doit être général, seulement la persévérance en tient lieu quelquefois.

Santé - Jai connu bien des savants dune mauvaise santé. Lorsquil en résulte, chez nous, la dispense du service auxiliaire, et que néanmoins la tête est bonne, la mauvaise santé est une chance de succès. Quand jétudiais le droit mon meilleur professeur, un très habile jurisconsulte, était estropié des jambes et des bras. Dans ce moment, à Genève, un des hommes les plus habiles, comme horticulteur et naturaliste, est un ancien pasteur chez lequel la vie sest retirée peu à peu des extrémités à la tête depuis 25 ans. Il ne peut porter sa nourriture à la bouche, mais il continue à lire des ouvrages scientifiques en 3 langues et dicte des traductions. Il fait porter son tronçon supérieur (le seul vivant) dans son jardin pour vérifier la suite dexpériences quil a ordonnées. Le premier de ces deux estropiés remarquables avait une grosse tête et ressemblait à Napoléon Ier; le second également énergique, en a un petite. Jai vu souvent les exercices de corps détourner des études ou détruire lhabitude dobserver soigneusement. Sans doute ils doivent être proportionnés au degré de force de chacun, mais pas au delà.

Practical business - Ceci est bien anglais ! Sur le Continent vous trouveriez une foule de savants qui nentendent rien aux alfaires, qui sen moquent, et dont la fortune diminue plutôt que daugmenter. La négligence des affaires est plus commune encore chez les hommes de lettres. Je crois bien quun savant réussit rarement sil na pas de lordre dans ses papiers, ses travaux et même dans les habitutes ordinaires de la vie, mais ce nest pas ce que vous appelez practical business (aptitude aux affaires commerciales, industrielles etc.).

Vos Réflexions p. 6... All tends to... divine, sont extrêmement justes, surtout le caractère anti-féminin! Je ne dirai pas pourquoi, ce serait offenser la plus belle moitié de notre espèce.

Jen dirai autant de lavantage dune instruction variée. Contrairement à lopinion des pédants de collèges, jestime avec vous que cest une grande source de curiosité et dautres avantages, qui se retrouvent ensuite dans la spécialité dune carrière. Seulement il ne faut pas continuer son éducation variée indéfiniment. Il faut savoir devenir spécial.

Dans lespoir de lire bientôt le volume projeté de vos recherches intéressantes, je suis toujours, mon cher Monsieur, votre très dévoué collègue

Alph. de Candolle.

4.10 1875/01/11. Alphonse de Candolle to Francis Galton

11 January 1875.

My dear sir,

If I was late in thanking you for sending your English men of Science volume, it's not that I neglected it.[14] On the contrary, I have read it twice with much interest and propose to quote it often if I do a second edition of my work on l'Histoire de Sciences. For the moment I am preparing another second edition, that of my Géographie botanique raisonnée, but I hope to return to the object that has occupied both of us.

[14] Galton 1874a.

We used two very different methods, which often led us to the same conclusions. That supports both. When the results are different it is probably because the peculiar conditions for the English are not those of most countries, and I believe that if the same questions were asked elsewhere, we would have quite different answers. It would not be easy to get those answers. Perhaps the individuals would be less truthful in the answers, sometimes deliberately and sometimes unwillingly. In England itself, where one is more truthful, some people are self-deluded. I could not help smiling as I read that some of your clergymen claim that they have not been upset by their opinions in scientific research. I have heard it said, with the same good faith, by some of our own, but when we talk to them about certain facts, they deny or evade them, and often they avoid certain sciences or some research. As Herbert Spencer puts it so well, in his Social Science, man is naturally inconsistent. Your clergymen seem more distant than ours from scientific tendencies, except Unitarians. Likewise, among us Trinitarians are often enemies of science and not Unitarians (liberal Protestants).

The geographical origin of your men of science (p. 20) may be due to the preponderance of cities in the shaded part.[15] The proportion favorable to Scotland, and unfavorable to Ireland, of which you speak elsewhere, was also very clear from my lists, and you explain, I think, with reason that Scots

[15] Of the map of the distribution of eminence, that is.

receive a better education. Geneva, Holland, Scotland and New England are or were branches of a vigorous intellectual tree planted by Calvin.

The quantity of learned single sons or first-born (p. 33) surprised me, but with a little thought it is understandable.[16]

You prefer statistical research, including fertility. That can be done in England better than on the Continent, better than in France, because the absence of children there is less often voluntary. I had no idea of the extent of this cause in the French provinces before having read a serious but not very good book by an old doctor who practices in the small town of Arbois: Bergeret, Des fraudes dans laccomplissement des fonctions génératrices, leurs causes, dangers, etc., 1 vol. 8vo Paris 1873, at L.B.Baillières, price 2 Fr. 50 c.[17] The author exaggerates, I strongly believe, the disadvantages of the abuses he highlights. Like all doctors, he sees only those who suffer from certain practices and does not think of all those who have not experienced any ill effects. But the debauchery he reports in very small towns and even in the rural communes of France explains the very low number of births and gives rise to singular reflections. Equal sharing between children is

[16] Candolle appears to say 'born' but 'first-born' makes the most sense.
[17] Des Fraudes Dans L'accomplissement Des Fonctions Génératrices Dangers Et Inconvénients Pour -Les- Individus, La Famille Et La Société Par L.-F.-E. Bergeret Médecin en Chef De L'hopital D'arbois (Jura). Paris J.-B. Baillière Et Fils Libraires De L'académie Impériale De Médecine Rue Hautefeuille, 19, Près Le Boulevard Saint-Germain. 1870

obviously a cause of immorality among parents and laziness among children. It is also, it is true, a cause of economy in families and total wealth. In England capital increases are divided between ever more numerous individuals; in France they remain accumulated in a static population.

I do not judge the conditions of health quite like you. It is possible that the sons of sturdy parents are less so when they become men of science, living in the cities, but they are nonetheless adapted to their environment and that is the essence. In Paris I often met the sons and grandsons of Parisians who did not look robust, but who bore the fatigue of urban life better than the countrymen. Many were remarkable for their intelligence and activity. By avoiding the cold and supporting themselves with good food these city dwellers, without having good muscles, live long and do an excellent job. They are naturalized—but for that it takes two generations at least. The country boys have often appeared to me to suffer from education in the city, as well as the sons of laborers in the country.

Your ideas about education are supported by the facts. They are excellent and I wish them success. The journal Nature often says the same things, but mixes them with assertions and prejudices that I can not always approve of. They seem to think that gentlemen are naturally inferior to the workers. We have both proved that the upper classes everywhere give a considerable proportion of eminent men. It would be useful, therefore, to favor those of this class who show some taste for the sciences, and it would be more profitable than to artificially elevate some of the lower classes by means of financial subsidies difficult to target properly. Your last volume completes the former admirably. I congratulate you and ask you to accept, my dear Sir, the assurance of all my devotion.

Alph. de Candolle.

11 Janvier 1875.

Mon cher Monsieur,

Si jai tardé à vous remercier de lenvoi de votre volume English men of Science, ce nest pas que je laie négligé. Au contraire je lai lu deux fois avec beaucoup dintérêt et me propose de le citer souvent si je fais une seconde édition de mon ouvrage sur lHistoire des Sciences. Pour le moment je prépare une autre seconde édition, celle de ma Géographie botanique raisonnée, mais jespère revenir ensuite à lobjet qui nous a tous deux occupés.

Nous avons employé deux méthodes bien différentes, qui nous ont souvent conduits aux mêmes déductions. Cest une preuve en faveur de toutes les deux. Lorsque les résultats diffèrent cest probablement que les conditions spéciales aux Anglais ne sont pas celles de la plupart des pays, et je crois bien en effet que si lon posait ailleurs les mêmes questions, on aurait des réponses assez souvent différentes. Il ne serait pas facile dobtenir ces réponses. Peut-être les individus seraient moins vrais dans les réponses, tantôt le voulant et tantôt sans le vouloir. En Angleterre même, où lon est plus véridique, certaines personnes se font des illusions. Je nai pas pu mempêcher de sourire en lisant que quelques-uns de vos ecclésiastiques prétendent navoir nullement été contrariés par leurs opinions dans les recherches scientifiques. Je lai entendu dire, avec la même bonne foi, à quelques-uns des nôtres, mais quand on leur parle de certains faits, ils les nient ou les éludent, et souvent ils évitent certaines sciences ou certaines recherches. Comme lexpose très bien Herbert Spencer, dans sa Social Science, lhomme est naturellement inconséquent. Vos ecclésiastiques paraissent plus éloignés que les nôtres des tendances scientifiques, exceptés les unitairiens. De même, chez nous les trinitairiens sont souvent ennemis de la science et non les unitairiens (protestants libéraux).

Lorigine géographique de vos hommes de science (p, 20) tient peut-être à la prépondérance des villes dans la partie ombrée. La proportion favorable à lEcosse et défavorable à lIrlande, dont vous parlez ailleurs, résultait aussi très clairement de mes listes, et vous expliquez, je crois, avec raison que les Ecossais reçoivent une éducation meilleure. Genève, la Hollande, lEcosse et la Nouvelle Angleterre sont ou étaient des branches dun arbre intellectuel vigoureux planté par Calvin.

La quantité de savants fils unique ou les[18] nés (p. 33) ma

[18] Reads 'lers' or 'Iers' but this cannot be correct.

étonné, mais avec un peu de réflexion cela se comprend.

Vous aimez les recherches statistiques, entre autres sur la fécondité. Elles peuvent se faire en Angleterre mieux que sur le Continent, mieux surtout quen France, parce que labsence denfants y est moins souvent volontaire. Je navais pas didée de létendue de cette cause dans les provinces françaises avant davoir lu un ouvrage sérieux, mais peu décent, dun vieux médecin qui pratique dans la petite ville dArbois : Bergeret, Des fraudes dans laccomplissement des fonctions génératrices, leurs causes, dangers, etc., 1 vol. 8vo Paris 1873, chez L.B.Baillières, prix 2 Fr. 50 c. Lauteur exagère je crois beaucoup les inconvénients des abus quil souligne. Comme tous les médecins il ne voit que ceux qui souffrent de certaines pratiques et ne pense pas à tous ceux qui nen ont pas éprouvé dinconvénient. Mais le libertinage quil signale dans de très petites villes et même dans les communes rurales de la France explique bien le nombre très faible des naissances et fait naître de singulières réflexions. Le partage égal entre les enfants est évidemment une cause dimmoralité chez les parents et de paresse chez les enfants. Cest aussi, il est vrai, une cause déconomie dans les familles et de richesse totale. En Angleterre les accroissements de capitaux se divisent entre des individus toujours plus nombreux ; en France ils restent accumulés dans une population stationnaire.

Je ne juge pas la condition de la santé tout à fait comme vous. Il est possible que les fils de parents robustes le soient moins lorsquils deviennent des hommes de science, vivant dans les villes, mais ils nen sont pas moins adaptés à leur milieu et cest lessentiel. A Paris jai rencontré souvent des fils et petit-fils de parisiens qui navaient pas lair robustes, mais qui supportaient mieux que les campagnards les fatigues de la vie urbaine. Plusieurs étaient remarquables par leur intelligence et leur activité. En se tenant à labri du froid et se soutenant par une bonne nourriture ces citadins, sans avoir de bons muscles, vivent longtemps et font un travail excellent. Ils sont naturalisés mais pour cela il faut deux générations au moins. Les fils de campagnards mont paru souffrir fréquemment de léducation à la ville de même que les fils douvriers dans la campagne.

Vos idées sur léducation résultent des faits. Elles sont excellentes et je leur souhaite du succès. Le journal Nature dit souvent les mêmes choses, mais il les mélange dassertions et de préventions que je ne puis pas toujours approuver. Il a lair de croire que les gentlemen sont naturellement inférieurs aux ouvriers. Nous avons prouvé lun et lautre que les classes supérieures donnent partout une proportion considérable dhommes éminents. Il serait donc utile de favoriser ceux de cette classe qui montrent quelque goût pour les sciences et ce serait plus profitable que délever artificiellement quelques individus de la classe inférieure, au moyen de subventions pécuniaires difficiles à bien placer. Votre dernier volume complète admirablement le premier. Je vous en félicite et vous prie dagréer, mon cher Monsieur, lassurance de tout mon dévouement.

Alph. de Candolle.

4.11 1875/03/05. Francis Galton to Alphonse de Candolle

42, Rutland Gate, London.
March 5/75

My Dear Sir

I have left your welcome letter unanswered for two months, being desirous of sending, which I now do, a daily expected copy of a paper read at the Anthropological Society, on a subject to which you directed attention in the "Histoire des Sciences et des Savants." It is on the probability of the decay of families by purely accidental causes. If you could persuade any of your Swiss mathematicians to pursue the subject, it would be very advantageous. What seems to be wanted now, is some simple function which approximately represents the distribution of children in families, and to work this with Watson's general formulae. You will see how complicated the problem is. I send the memoir by book post.[19]

[19] The extinction of surnames problem was first posed in Galton 1873c. The Rev. Henry William Watson, a mathematician who had been second wrangler and Smith's prizeman at Trinity College in 1850, proposed an approximate solution, which was published as Galton

ingenious little instrument of Swiss invention and manufacture, 'Amsler's Planimeter'. I have had it largely tested and employed at the Meteorological Office of England (of which I am one of the managing Committee) with perfect success. A full description of its employment will be found in our Meteor Office 'Quarterly Weather Reports' of last year which are in the Geneva Observatory.

If the desire be, to try sums of the squares of excess of temperature, or of any other function, the same method of summation is of course equally applicable.

Pray excuse my prolixity; I write on the chance that our meteorological experience of rapid methods for avoiding tedious computation may prove of service in your further inquiries.

I am writing from Switzerland, from Thun, whither my wife and I are shortly going towards, the Lake of Geneva; I had your pamphlet sent to me here. Should I be in the neighbourhood of Geneva, I will certainly do myself the pleasure of calling at your house in the hope (I fear a faint one at this season) of finding you at home.

Yours very faithfully,
Francis Galton.

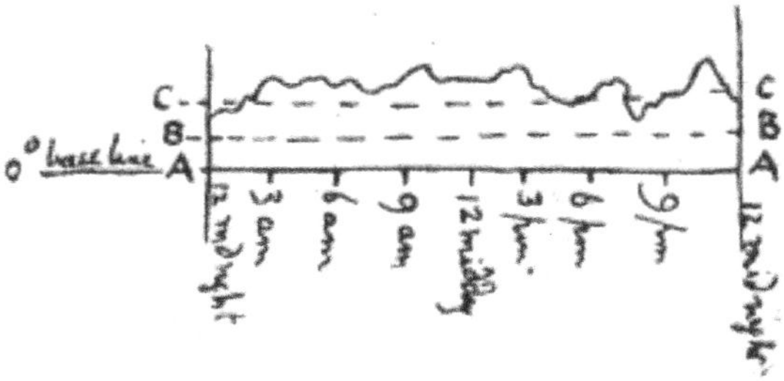

Figure 4.8: 1875-07-22 FG to Candolle, Thermogram.

4.13 1876/05/24. Alphonse de Candolle to Francis Galton

May 24, 1876.

My dear Sir,

I'm sending you (enclosed) an article about your interesting pamphlet about twins. It has just been published in the Archives of Physical and Natural Sciences of May 1876 which is published here. I have added here and there some reflections to emphasize the interest of your researches. I regret that your departure from Switzerland coincided with my excursion in the Engadine, where my health forces me to go during the heat. Accept, I pray you, the assurance of my eager greetings.

Alph. de Candolle.

24 mai 1876.

Mon cher Monsieur,

Je vous envoie (sous bandes) un article sur votre intéressant opuscule relatif aux jumeaux. Il vient de paraître dans les Archives des Sciences physiques et naturelles de mai 1876 qui se publie ici. Jai ajouté çà et là quelques réflexions pour montrer mieux lintérêt de vos recherches. Je regrette que votre départ de la Suisse ait coïncidé avec mon excursion dans lEngadine, où ma santé moblige à aller pendant les grandes chaleurs. Agréez, je vous prie, lassurance de mes salutations empressées.

Alph. de Candolle.

4.14 1879/11/08. Alphonse de Candolle to Francis Galton

8 Nov. 1879.

My dear Sir,

I read with pleasure the pamphlets that you kindly sent me.[20] The composite photographs seemed curious to me. They will probably be used widely and for various researches in medicine, physiology, etc. We can also expect some useful follow-ups in the arts. I propose, for this reason, to show your generic images in a session of our Society of Arts where there are always designers, photographs, etc. If you publish other essays I will be very obliged to be sent them by you before the next January.

[20] Galton 1878, 1879a,b.

The appearance of your criminals confirms what I read yesterday in a scientific article signed Parville in Temps, on the skulls of 36 French assassins. Their skulls were wide from one temple to another.

The psychometric observations of which you describe the results seem to me very difficult to conduct. However, I know from experience that we must wrestle with facts of this nature. I have also observed similar things during the night, and what I have concluded supports your opinion by quite interesting comparisons.

For two years, at the age of 71 and 72, I tried to grasp, at the moment of waking, the nature of my dreams. With a firm will we may succeed. The details are forgotten very quickly, but we can recall it if the dream relates to old or new things and things that we had taken care of the day before or which we had not thought for a long time. Here is what I found:

1. Very often in the evening I spoke about some person or object, which served as a starting point for a dream. Sometimes it had been spoken of in my presence, or I had read aloud the name of the person or object. Rarely did reading by eye produce this consequence. The name or the word had become the source of some involuntary association of ideas, as in your observations, but the links were meandering and often absurd. Conversations or reading of a few days earlier, even I believe 24 hours, did not lead to dreams.

2. (and this even more certain): Never have I dreamed

of things that caused me, either recently or formerly, strong anxieties or strong emotion. Although my career has been fairly uniform, I have had sorrows, I have had problems that prevented me from sleeping. I witnessed revolutionary scenes that annoyed me in the highest degree. My life has been in peril several times in steep ascents of mountains, etc. These events have never presented themselves in my dreams, nor have they been very lively old pleasures.

3. I have dreamed of people who have been dead for a long time, but almost always I remembered that I had spoken about it in the evening, or a similar connection of ideas had existed. For example, I often dreamed of being with my father, who died in 1841: he appeared before me, to talk with me about scientific matters with a lot of construing, but I work every day in my father's library; I consult his works, etc. It's in the usual stream of my ideas.

All this confirms, from another source, your thoughts on page 7.

The ideas we have at night, when we are awake and we can not fall asleep, also confirm what you say. In this state one has (at least I do) two or three very specific ideas that are common. These are ideas that have been bothering you for a few days: probably a concern about one of your relatives, a trial, etc. or some difficult letter to write, some speech to make, etc. As we see nothing [else while asleep], other causes of ideas do not exist and those which dominate you have an extraordinary vitality. A little nervous agitation that prevents sleep increases the vivacity of nocturnal ideas. I was so often struck with their sharpness that I had made myself a slate, with a moving ruler, to be able to write in my bed certain phrases, certain divisions of a subject, which appeared to me suddenly and that the next day I could not find again. The apparatus is not very convenient to use [in the dark], so sometimes I turn on a light to record what has just appeared to me in these nights of insomnia. For a writer, the absence of distractions seems to me an essential cause of success. I do not understand at all those who study or write while walking in the countryside.

There, my dear sir, are some observations of which you will do what you wish. I do not intend to publish them. If you wish to speak of it in some note, I have no objection, without, however, asking you to do so.

Accept, I pray you, the assurance of my most devoted greetings.

Alph. of Candolle.

8 nov.1879.

Mon cher Monsieur,

Jai lu avec beaucoup de plaisir les opuscules que vous avez bien voulu menvoyer. Les photographies cumulées mont paru curieuses. Elles serviront probablement dans plusieurs cas et pour diverses recherches de médecine, physiologie, etc. On peut en inférer aussi certaines conséquences utiles dans les arts. Je me propose, par ce motif, de montrer vos generic images dans une séance de notre Société des arts où se trouvent toujours des dessinateurs, des photographies, etc. Si vous publiez dautres essais je vous serai très obligé de me les envoyer avant le mois de janvier prochain.

Laspect de vos criminels confirme ce que je lisais hier dans un article scientifique signé de Parville dans le Temps, sur les crânes de 36 assassins français. Leurs crânes étaient larges dune tempe à lautre.

Les observations psychométriques dont vous décrivez les résultats me semblent bien difficiles à faire. Cependant je sais, par expérience, quon arrive en se donnant de la peine à constater des faits de cette nature. Jai aussi observé des choses analogues pendant la nuit, et ce que jai conclu appuie votre opinion par des rapprochements assez intéressants.

Pendant deux années, à lâge de 71 et 72 ans, je me suis appliqué à saisir au moment du réveil la nature des mes rêves. Avec une ferme volonté on y parvient. Les détails soublient très vite, mais on peut notre dans sa tête si le rêve se rapporte à des choses anciennes ou récentes et à des choses dont on sétait occupé la veille ou auxquelles on navait pas pensé depuis longtemps. Voici ce que jai trouvé :

1. Très souvent dans la soirée javais parlé de quelque personne ou objet, qui a servi de point de départ à un rêve. Quelquefois on en avait parlé devant moi, ou javais lu à

haute voix le nom ou celui de lobjet. Rarement une lecture des yeux produisait cette conséquence. Le nom ou le mot était devenu source de quelque association involontaire didées, comme dans vos observations, mais les déductions étaient errantes et souvent absurdes. Des conversations ou lectures de quelques jours antérieures, même je crois de 24 heures, ne conduisaient pas à des rêves.

2. (et ceci encore plus certain) : jamais je nai rêvé à des choses qui mavaient causé ou récemment ou autrefois de vives inquiétudes ou une vive émotion. Quoique ma carrière ait été assez uniforme jai éprouvé des chagrins, jai eu des soucis qui mempêchaient de dormir. Jai assisté à des scènes révolutionnaires qui mirritaient au plus haut degré. Ma vie a été exposée plusieurs fois dans des courses de montagne, etc. Or ces événements ne se sont jamais présentés dans mes rêves, non plus que des plaisirs anciens très vifs.

3. Jai rêvé à des personnes mortes depuis longtemps, mais presque toujours je me suis souvenu que jen avais parlé dans la soirée, ou quune liaison didées analogue avait existé. Par exemple je rêvais souvent être avec mon père, mort en 1841 : il me semblait le voir, causer avec lui, sur des affaires scientifiques avec beaucoup de suite et de raison, mais je travaille tous les jours dans la bibliothèque de mon père; je consulte ses ouvrages, etc. Cest dans le courant habituel de mes idées.

Tout cela confirme, par une autre voix, vos réflexions de

la page 7.

Les idées quon a la nuit, quand on est réveillé et quon ne peut pas sendormir, confirment également ce que vous dites. Dans cet état on a (du moins moi) deux ou trois idées très précises qui reviennent couramment. Ce sont des idées qui vous préoccupent depuis quelques jours : probablement une inquiétude sur un de vos proches, un procès, etc. ou quelque lettre difficile à rédiger, quelque discours à faire, etc. Comme on ne voit rien, dautres causes didées nexistent pas et celles qui vous dominent ont une force extraordinaire. Un peu dagitation nerveuse qui empêche de dormir augmente cette vivacité des idées nocturnes. Jai été si souvent frappé de leur netteté que je métais fait fabriquer une ardoise, avec règle mobile, pour pouvoir écrire dans mon lit certaines phrases, certaines divisions dun sujet, qui mapparaissaient tout-à-coup et que le lendemain je ne pouvais plus retrouver. Lappareil nest pas assez commode pour lemployer, aussi quelquefois jallume une lumière pour noter de qui vient de mapparaître dans ces nuits dinsomnie. Pour un écrivain labsence de sujets de distraction me paraît une cause essentielle de succès. Je ne comprends pas du tout ceux qui étudient ou rédigent en se promenant dans la campagne.

Voilà, mon cher Monsieur, des observations dont vous ferez ce que vous voudrez. Je nai pas lintention de les publier. Si vous voulez en parler dans quelque note je nai pas dobjection, sans cependant vous le demander.

Agréez, je vous prie, lassurance de mes salutations les plus dévouées.

Alph. de Candolle.

4.15 1880/01/19. Francis Galton to Alphonse de Candolle

Jan. 19/80

Dear Sir,

I take the opportunity of sending for your acceptance a copy of "Nature" containing an article on a rather curious inquiry "Visualised Numerals" (which you might care to glance at) to thank you heartily for a kind letter full of suggestions, hints and information which you wrote me many weeks ago.[21] It will be of great use to me.

Yours faithfully
Francis Galton.

4.16 1880/04/09. Francis Galton to Alphonse de Candolle

Geneva April 9, 1880

My dear sir

[21] Galton 1880b. See also Galton 1880c,d.

Since you sent me your interesting pamphlets on associations of ideas, normal and abnormal,[22] I looked around to see if they are as frequent as you think. To that end I have spoken of your observations to a few persons, and I raised it at a meeting of the members of the Society of the Arts, at which there were artists, scientists, industrialists, and persons of various professions or conditions.

I have only identified three or four people who are accustomed to associating ideas that have no natural connection, such as the shape of the letters with a color, the numerals with a certain spatial arrangement, and so on. Only one of these people seemed to me to deserve great confidence because he is an engineer, a mathematician, an accurate and truthful person. Another was a lady, not without ability to observe, but I have often found that the imagination plays a greater role in women than in men, not to mention the desire to make an impression in a parlour conversation. The rest were hard to interpret. In spite of that I think that the disposition to associate things which have no real relation is quite ready in men, and that one can especially develop it by following some processes, which are nevertheless hardly worthy of encouragement. Enthusiasts and charlatans who love mnemonics or who exploit them, are the proof. Associate three or four times the succession of the Kings of England, to successive panels of a room, and the connection will then easily be by rote. The

[22] Galton 1879a, 1880a.

more one indulges in this kind of exercise, apt to deform the soul, the more you become proficient. It's exactly like playing puns.

Poets associate words, not by their meanings, but by their final letters. As there have always been poets, especially in ancient times, man must have an unfortunate disposition to false and fictitious associations. The sciences counteract this trend, but they only act on a small number of men and an even smaller number of women. Most prejudices are founded on ill-conceived associations of ideas, and we see that ridiculous prejudices prevail or at least exist in civilized countries. I thought I saw that in England people feared Fridays, salt shakers, etc. as much and perhaps even more so than in many countries. In a family of naturalists like ours, where we have never sought out associations, and retained only true instances, these anomalies seem very singular. I do not detect the slightest trace of this in myself.

Receive, my dear sir, the assurance of my most eager greetings.

Alph. de Candolle

Genève 9 avril 1880

Mon cher Monsieur

Depuis que vous mavez envoyé vos intéressants opuscules sur les associations didées, naturelles ou anormales, jai cherché à voir autour de moi si ces dernières sont aussi fréquentes

que vous le pensez. Dans ce but jai parlé de vos observations à quelques personnes et jen ai entretenu une assemblée des membres de la Société des Arts dans la quelle se trouvaient des artistes, des savants, des industriels et des personnes de diverses professions ou conditions.

Je nai constaté que trois ou quatre personnes ayant lhabitude dassocier des idées qui nont aucun rapport naturel, comme la forme des lettres avec une couleur, les chiffres avec un arrangement matériel déterminé, etc. Une seule de ces personnes ma paru mériter une grande confiance, parce que cest un ingénieur, mathématicien, exact et véridique. Une autre était une dame, non sans capacité dobservation, mais il mest arrivé souvent de constater que limagination joue un plus grand rôle chez les femmes que chez les hommes, sans parler du désir de paraitre ou de faire de leffet dans une conversation de salon. Enfin les autres personnes naffirmaient pas clairement ce quelles disaient. Malgré cela je pense que la disposition à associer des choses qui nont pas de rapports réels est assez facile chez lhomme, et surtout quon peut la développer par certains procédés, peu dignes dencouragement. Les amateurs et charlatans qui aiment la mnémotechnie ou qui lexploitent, en sont la preuve. Associez trois ou quatre fois la succession des rois dAngleterre, à des panneaux successifs dune salle, et la liaison se fera ensuite aisément par mémoire, Plus on se livre à ce genre dexercice, propre à fausser lesprit, plus on y devient apte. Cest exactement comme de faire des jeux de

mots.

Les poètes associent les mots, non par leurs sens, mais par leurs lettres finales, Comme il y a eu toujours des poètes, surtout dans les temps anciens, il faut que lhomme ait une disposition regrettable aux associations fausses et factices. Les sciences combattent cette tendance, mais elles nagissent que sur un petit nombre dhommes et un nombre encore plus petit de femmes. La plupart des préjugés tiennent à des associations didées mal conçues, et nous voyons que les préjugés ridicules règnent ou du moins existent dans les pays civilisés. Jai cru voir quen Angleterre on craint le vendredi, les salières renversées, etc. autant et plus peut-être que dans plusieurs pays. Dans une famille de naturalistes, comme la nôtre, où lon na jamais cherché et retenu que des associations vraies, ces anomalies paraissent très singulières. Je nen sens pas en moi le plus petit indice.

Recevez, mon cher Monsieur, Lassurance de mes salutations les plus empressées.

Alph. de Candolle

4.17 1880/04/11. Francis Galton to Alphonse de Candolle

42, Rutland Gate, London.
April 11/80.

My dear Sir,

Thank you very much for the kind efforts you have made to procure me information about the visualised numerals. They have caused M. Achard to send me his numeral forms and some interesting accompanying remarks, which I have added to my collection.[23] Perhaps the enclosed reprint may interest you, if you have not by chance already seen it, as it gives a recent account of the facts and some remarks upon them which I think you will find to answer in part the very reasonable doubts you suggest in your kind letter, which I received last night.

I do not think these forms of any value to those who see them, nor that they should be cultivated, but they strike me as exceedingly curious and instructive survivals of the earliest mental processes of a child. They are specially interesting because of the reasons given in the enclosed reprint, they have been invented by the child himself, but I will not write what you will more rapidly read in print.

It has been a most amusing but somewhat discouraging experience to find how very many wise men are, as it were, vexed and put out by finding that other people have real undeniable gifts that they do not themselves possess a vestige of, and are inclined in consequence to discredit the inquiry. M. Antoine d'Abbadie who sees these number-forms clearly, kindly questioned for me several of his colleagues of the Académie

[23] Arthur Achard (1836–1931), a civil engineer in Geneva who published widely in technical journals.

des Sciences and came to just the same result that I did. It was therefore with some wicked feeling of triumph that I collected and marched off with to the evening meeting of the Anthropological Institute, six good men including persons well known to science, who were prepared to describe their number-forms and who did so very effectively. I am now busy on a more generally interesting part of the subject of Mental Imagery. Believe me with thanks, and warm acknowledgement of the kind interest you have so often shown in my work,

Very faithfully yours, Francis Galton.

I am afraid our mutual friend Mr Bentham[24] has felt the gloom and severity of this past winter, for he does not look well, and complains about himself.

4.18 1882/06/05. Francis Galton to Alphonse de Candolle

42, Rutland Gate, London.
June 5/82.

My Dear Sir,

Thank you much for your interesting brochure on Ch. Darwin, analysing the causes that contributed to his success.[25]

[24] George Bentham (1800–1884) the botanist.
[25] 'Darwin aieul, considéré au point de vue des causes de son succès et l'importance de ses travaux' [Darwin, considered from the point of

It has been a great satisfaction in all our grief at his loss, to witness the wide recognition of the value of his work. He certainly, as you say, appeared at a moment when the public mind was ripe to receive his views. I can truly say for my part that I was groaning under the intellectual burden of the old teleology, that my intellect rebelled against it, but that I saw no way out of it till Darwin's 'Origin of Species' emancipated me. Let me, while fully agreeing with the views expressed in the pamphlet in all important particulars, supply a few minor criticisms which it might be well to mention.

(1) As to the pecuniary fortune of Darwin, I think the phrases "moyenne pour l'Angleterre etc."—"la maison modeste..." (pp. 12-13)[26] hardly convey the right idea. I should think that his fortune was much more considerable—say upwards of £5000 a year, before his brother's death in 1881, and subsequently larger. The house was maintained in thoroughly substantial and costly comfort, but when the particulars of the Will are published, which I suppose they soon will be, we shall know.[27]

(2) "Les descendants du poete phys[iologiste] p. 12...ont lu certainement de bonne heure les ouvrages de leur aieul."[28]

---

view of the causes of his success and the importance of his work] tiré des *Archives des Sciences de la Bibliothèque Universelle*, Tome VII. May 1882.

[26] "average for England etc.—the modest house ...".

[27] Darwin's personal estate was valued at over £146, 911.

[28] "The descendants of the poet-physician p. 12 ... definitely read the works of their ancestor early on." In reference to Erasmus Darwin.

I am almost certain of the contrary in every case except Ch. Darwin, (and I doubt in his case whether he had). To myself the florid and now ridiculed poetry was and is intolerable and the speculative physiology repellent. I had often taken up the books and could never get on with them. Canning's parody "the loves of the triangles" quite killed poor Dr Darwin's reputation. It just hit the mood of the moment and though my mother never wearied of talking of him, his life was to me like a fable only half believed in. That much the same was the case with some of Charles Darwin's sons, I can, I think, affirm.

(3) George Darwin "deja connu par de bons memoires de statistique" (p. 13).[29] Probably you may not know his present very high position as a mathematical astronomer, who has revealed the past history of the planetary system, in a most unexpected way. His works are spoken of in the presidential address of the Royal Society etc. as massive works. They are only slowly becoming known, being exceedingly laborious mathematical work of a kind that is within the practice of very few men indeed, but by them cordially recognised as commensurate in originality and importance with that of Laplace. His calculations depend on the "viscosity" of all solid bodies on the yielding of their substance to a tidal action, and most unexpected results came out, which bind under one scheme a large variety of astronomical phenomena.

[29] "Already left a good impression on statistics".

When I received your pamphlet, it so happened that your name had just been on my lips in respect to quite another matter, in which you were at one time much interested and which is now being taken up here. It is a question of cumulative temperature on vegetation. I have been since the beginning one of the members of the council to whom a large annual grant is entrusted by Government to carry on the systems of forecasts in land and ocean meteorology and we are endeavouring to give weekly data that may be of direct use to agriculturists. In reply to questions that we circulated as to the best form for that purpose, frequent mention was made of the cumulative values of heat. We have accordingly been investigating the probability of calculating these values in units of 'day-degrees' viz. (1) cumulative effect of heat derived from 1°Fahr. of temperature acting during 24 hours, or of (2) acting during 12 hours, and so on. The result is that it is quite feasible to do so, with fair approximation, on the data of registered days' maxima and minima, and accepting any arbitrary base-line above which the accumulative temperature is to be reckoned. We can easily give 2 or more slices of the diurnal curve; that is to say the cumulative values between 3 or more arbitrary temperatures.

My colleagues ask me to inquire of you whether you happen lately to have again attended to the subject or whether you have any suggestions to make that might help us, in addition to what you have already published and which we find to be thoroughly appreciated by some of our correspondents?

It is rather out of our line to do so, but we might perhaps, if it were thought essential, get experiments made on the cumulative effects of temperature on some forms of vegetation—say the cereals—but probably sufficient information for our purpose already exists. We can measure cumulative effects of sunshine, rain, and temperature and could measure that of evaporation under any one definite condition, but it is a question whether the latter limitation would not render the results of little general service. I should be greatly obliged for a reply to the above question. Believe me, my dear Sir,

very faithfully yours,
Francis Galton.

4.19 1884/01/14. Alphonse de Candolle to Francis Galton

January 14, 1884.

My dear Sir,

I am preparing a second edition of my History of Science and Scholarship, which has been out of print for a long time and which the booksellers request from me. For that I am making extensive use of your English men of Science and the recent volume of Inquiries into human faculties which contains many curious articles. We practice the same method, that of observing, and when we can, counting to compare, therefore we must support each other and we stand to gain little in opposition.

Let me ask you some information about English scholars.

Could you tell me what were the positions or professions of the fathers of the famous zoologist Owen, newly created H. C. B., Sir George Airy and Sir George Wheatstone, I could not ascertain that from the biographical dictionaries at hand.

I presume that Sir William Thomson, born in Belfast, the son of a maths teacher, was a Protestant of a Scottish or English family settled in Ireland. Is that correct?

The character of your illustrious cousin Charles Darwin is so honorable, so eminent in many respects, that I would like to know about him some details of even secondary value. For example, did he have a natural disposition to drawing? and to music? Nothing suggests it in his works.

I do not know whether to ascribe to him a strong imagination. Many people reproached him because they did not understand the value of his observations and deductions, and it pleased them to say that he indulged in pure hypotheses. For me, who quickly recognized the wisdom of his mind, and his prudence, I do not know whether these qualities occupied the whole place of the imagination, or if it must be conceded that along with much vigor of reasoning he had much imagination.

The eldest of my grandchildren, Raymond de Candolle, born an Englishman, raised at Rugby, and having just entered

Trinity College, Cambridge, sees the numbers arranged in series and certain figures more apparent than others in his mind.[30] This is the only case of this kind in my family.

Always, my dear sir, your very devoted

Alph. de Candolle.

14 Janvier 1884.

Mon cher Monsieur,

Je prépare une seconde édition de mon volume de Y Histoire des sciences et des savants qui est épuisé depuis longtemps et que les libraires me demandent. Pour cela je fais grand usage de vos English men of Science et du volume récent des Inquiries into human faculties qui contient beaucoup darticles curieux. Nous suivons la même méthode, celle dobserver, et quand on le peut, de compter pour comparer, par conséquent nous devons nous appuyer lun lautre et nous risquons bien peu dêtre en opposition.

Permettez-moi de vous demander quelques informations sur des savants anglais.

Pourriez-vous me dire quelles étaient les positions ou professions des pères du célèbre zoologiste Owen, nouvellement créé H. C. B., de Sir George Airy et de Sir George Wheatstone, Je nai pas pu le savoir daprès les dictionnaires biographiques à ma portée.

[30] Raymond de Candolle (1864-1935) later served in the British Army as an engineer.

Je présume que Sir William Thomson, né à Belfast, fils dun prof, de mathématiques, était un protestant, dune famille écossaise ou anglaise établie en Irlande. Est-ce exact?

Le caractère de votre illustre cousin Charles Darwin est si honorable, si éminent sous plusieurs rapports, que jaimerais connaître sur lui certains détails dune valeur même secondaire. Par exemple, avait-il une disposition naturelle aux arts du dessin? et à la musique? Rien ne lindique dans ses ouvrages.

Je ne sais pas sil faut lui attribuer une imagination forte. Beaucoup de personnes le lui reprochaient, parce quelles ne comprenaient pas la valeur de ses observations et déductions, et quil leur plaisait de dire quil se livrait à de pures hypothèses. Pour moi qui ai reconnu très vite la sagesse de son esprit, et sa prudence, je ne sais pas si ces qualités avaient occupé la place entière de limagination, ou sil faut admettre que même avec beaucoup de vigueur de raisonnement il avait beaucoup dimagination.

Laîné de mes petits-fils, Raymond de Candolle, né Anglais, élevé à Rugby et qui vient dentrer à Trinity College, Cambridge, voit les chiffres disposés en séries et certains chiffres plus apparents que dautres dans son esprit. Cest le seul cas de ce genre dans ma famille.

Toujours, mon cher Monsieur, votre très dévoué

Alph. de Candolle.

4.20 1884/01/27. Francis Galton to Alphonse de Candolle

42, Rutland Gate, London.
Jan. 27/84.

My Dear Sir,

I delayed answering until I had an opportunity of talking over the questions you put about Darwin, with his very intelligent daughter.

He did not draw, he had not a good ear for music, but was much affected by it, sometimes to tears. He had naturally, (excuse the word which I know you detest! but I mean 'innately') a very emotional disposition, which was repressed by his habits of hard thinking, but always ready to burst out. Thus his delight in the scenery of a tour about the English Lakes a few years ago, had all the freshness and eagerness of that of a boy. However his nature could not be called aesthetic. As regards imagination I hardly know whether I understand the word in your sense, nor indeed if I have any definition of my own. I know that his faculty of mental imagery was once vivid and had become diminished, both from what he distinctly told me and from corroborative evidence. But that he ever was deceived by imagination I should think most unlikely, as he was so remarkably veracious.

He may be said to have studied veracity as the highest of arts. If imagination is cited in the sense of living in an ideal world of day dreams and poetry, I understand he was very fond of poetry as a boy but his interest in it faded by disuse. His scientific imagination in the sense of the power of envisaging abstract ideas, and living among them, and interesting himself with them was obviously great, on the evidence of his works. I am sorry not to be able to give you at present much information about the other men of whom you ask.

Wheatstone was an artisan, a mere workman originally. He took much interest in my inquiries and helped me in any way I asked, except as to his own history.

Airy promised to send me details, but eventually did not. His parents on both sides came of substantial farmers, solid men, of local notoriety. A certain disposition to dominate in argument is a strongly marked hereditary characteristic on the maternal (Biddell) side.

Owen I forget at the moment; they were low rather than high-middle-class. In a few days I shall be in the way of reviving recollections and will write again.

Thomson (Sir W.) I think you are quite right but here also I will write again.

Excuse this imperfect letter, but I am on the point of going to the country for a while and thought it best to write before going rather than after my return. I am very glad indeed that you are about to issue a new edition of your admirable volume. Let me say about the Darwin family that 4 of the 5 sons have achieved a very considerable reputation here.

George the Plumian Professor at Cambridge is looked upon as one of the ablest of the rising men in mathematical physics. He has made a great mark already and is rapidly rising in repute.

Frank who lectures at Cambridge on Botany was invited to be a candidate for the Professorship of Botany at Oxford, with a certainty of election, but for domestic reasons he refused.

Horace has set up, in conjunction with a friend, a laboratory at Cambridge for the manufacture of high class scientific instruments. He is most ingenious as a mechanical originator.

Leonard the Captain of Engineers, is one of the most scientific of his standing in that scientific corps. In the entrance examination he was first of all the candidates.

Thank you for telling me of your son's "number-form." I feel a little wicked delight at the fact occurring in your own family, because I recollected that you were at first somewhat sceptical of the reality of that curious tendency.

You will be glad to hear that we have begun at the Meteorological Office to publish accumulated temperatures in units of "day-degrees," counting (1) from Jan. 1, (2) from the first of the current week. General Strachey has worked out a beautiful method of obtaining them approximately, from the data of the daily maxima and minima, and a monthly (and probably a local) constant. I will tell them at the office to send you one of our new sheets as a specimen. It will give me the greatest pleasure if at any time I can be of service to you in obtaining information.

Yours very sincerely,
Francis Galton.

Our friend Mr Bentham continues very weak, but he has no organic malady.

4.21 1884/10/17. Francis Galton to Alphonse de Candolle

42, Rutland Gate, London.
Oct. 17/84.

My Dear Sir,

I have read and re-read your new edition of the "Histoire des Sciences" with great interest and instruction, and trust you will appreciate my attention to even the briefest criticism by the improved handwriting of this letter in deference to what you justly say (and said before) at the bottom of p. 541. It is very singular bow closely in many respects, our lines of inquiry run side by side. I shall be very curious indeed to see how far my own data will confirm yours in the 'nouvelles recherches,' but doubt much whether they will show the effect of heredity to be so strong, especially, for example, in myopism. Your appraisement of the several faculties, and selection of the faculties most convenient to be appraised, falls in very closely with an effort I lately indicated in the "Fortnightly" and am now making, to find out the best data by which the appraisement may be swiftly and fairly made. It has struck me that the masters and mistresses of schools might be able to indicate some often recurrent events in ordinary school life, which evoke different conduct in different children, and that by statistics of their conduct on these occasions, some fair guide to their habits and therefore to their character at the time being might be obtained. I should be greatly obliged for any hints that your experience may have suggested, how to appraise these qualities.

I venture to send you the "Fortnightly" of which I speak, not only on account of that article, but because of a very curious one in it about the Jews, by L. Wolf, which I feel sure will interest you in connection with your remarks on p. 174.[31] I think however, that Mr Wolf overstates his case. We have arranged to talk the matter over and he will show me his data. It strikes me that the Jews are specialised for a parasitical existence upon other nations, and that there is need of evidence that they are capable of fulfilling the varied duties of a civilised nation by themselves.

I see that you still adhere to your view of the influence of the parental feelings at the time of conception, on the child. Could not that be experimentally tested upon such animals

[31] Lucien Wolf (1857–1930) 'What is Judaism? A Question of Today.' *Fortnightly Review* New Series 36 (Aug. 1884), 237–256.

as rabbits? and possibly upon guinea pigs? The cause of fear might be the exhibition of a weasel or ferret. I cannot conclude without expressing my sincere pleasure at the way in which you have spoken of my inquiries. It is one of the pleasantest feelings to know that one is in intellectual sympathy with others.

Believe me very sincerely yours,

Francis Galton.

4.22 1884/10/27. Alphonse de Candolle to Francis Galton

Oct. 27, 1884.

My dear sir,

I am very happy to know that you approve my latest research on heredity. No sooner, however, was it published than I noticed other considerations which would merit examination and advance our knowledge. I hope that they will be in the next volume.

There is no doubt that schoolteachers could make interesting observations about children, but that requires observation, and time is often lacking for people so tired by their duties. If they are ordered to make such and such an observation it is worse than if it comes voluntarily.

The schools that could be most widely used are polytechnic, normal, naval, etc. in France. One enters at a certain age after examinations which classify individuals. Students are subject to the same diet of food, etc. Then we classify them on leaving. One could thus see whether, under equal conditions, the most skilful were the less robust, the blond or the brown, the smokers or the nonsmokers, the sons of educated men or others, &c. It would be necessary in one of these schools that a doctor or one of the administrators had the idea of making these comparisons; that is the difficult thing, although two years are sufficient.

Your article "Measurement of character" has the merit of provoking reflections and may give rise to valuable observations. Thank you for sending it to me!

Mr. Wolff's[32] article on the Jews seems a little too flattering to them. If certain prescriptions of Talmud are good for health, it seems to me very doubtful whether they are sufficiently observed by the faithful, as the author claims. The injunctions of Jews and Muslims on ablutions did not make them generally clean. They are less clean than many Christians, whom religion orders to despise their bodies, rather to think especially of the soul and of the future life. I can not believe that the separation of the spouses during periods as long as those of which Mr. Wolff speaks, are really observed, especially the early marriages, and if we confirm them we must admit there is a palliative in polygamy. The ancient Jews were regularly polygamous; the servants have probably served

[32] Lucien Wolf. See FG to AC, 1884/10/17.

as an alternative to the legitimate woman and today what is happening to the Israelites? I would like other testimonials than those of Mr Wolff, for example those of doctors. There are many Jewish actresses and models for painters. The usual foresight of the race and their early marriages are probably more important in regulating morals than the prescriptions of the religious law. Mr. Wolff's article has taught me many things that I did not know and his reflections are often very accurate.

You are right in saying that Jews are adapted to parasitic life. It's a good characterization of the facts. It must be said that they were forced to this exceptional life. If the difficulties were completely removed for them, they might change. D'Israeli was a statesman equal to many of the most distinguished.

It was a professor of Jewish physiology who proposed collaborating with me on experiments on the effect of alcoholism, fear, etc. on the results of rabbit procreation, or guinea pigs or dogs. I admit that these kinds of experiments are repugnant to me, but I think that they would prove what I suppose: that the momentary state of the parents influences the offspring. We have seen it many times for alcoholism and I have just read in a very serious review (Revue d'hygiène, October 1884, p. 875) the following:

> "Our friend, Prof. Layet, of Bordeaux, had to report to the Lattaie Health Congress on the vol-

> untary restriction of the birth rate from the point of view of its humanitarian and social consequences. According to him: 'From the moral point of view, it favors illegitimacy, the 9 departments with the least legitimate births also have the highest coefficient of illegitimacy ... The incomplete fulfillment of a function perverts the excitations instead of extinguishing them. The habit of restriction brings about a disturbance of the nervous system of the spouses; children born by mistake in these conditions feel the nervous disturbance that presided over their conception. The insane are more numerous in the departments where the husbands have the least number of children.'
>
> Mr. Lunier, of Paris, is convinced of the influence of abstinence on the development of the neurotic state, particularly of hysteria and lumbar neuralgia in women, and mentions many new instances. The suppression of the right of reproduction makes the disease spread like an oil stain in the countryside."

Certainly in countries, like France, where the law ensures equal inheritance between children, many women and husbands are afraid of a 3rd, 4th or 5th pregnancy. The effect of this fear on some children is likely. Here I am on a very scabrous subject. It is better to stop and pray to you, my dear sir, to believe me always your very devoted

Alph. of Candolle.

27 Oct. 1884.

Mon cher Monsieur,

Je suis très heureux de penser que vous approuvez mes dernières recherches sur lhérédité. A peine cependant elles étaient publiées que je voyais dautres considérations qui mériteraient examen et avanceraient nos connaissances. Jespère quelles se trouveront dans votre prochain volume.

Il ny a pas de doute que les maîtres et maîtresses décoles pourraient faire des observations intéressantes sur les enfants, mais il faut vouloir observer, et le temps manque souvent à des personnes aussi fatiguées par leurs fonctions. Si on leur ordonne de faire telle ou telle observation elle se fait moins bien que si cela vient de leur propre volonté.

Les écoles dont on pourrait tirer le plus grand parti sont par exemple les écoles polytechniques, normale, navale, etc. en France. On y entre, à un âge déterminé, après des examens qui classent les individus. Les élèves sont soumis au même régime de nourriture, etc. Ensuite on les classe à la sortie. On pourrait donc constater si, dans des conditions égales, les plus habiles ont été les pins on les moins robustes, les blonds ou les bruns, les fumeurs ou les non-fumeurs, les fils dhommes instruits ou dautres, etc. Il faudrait dans une de ces écoles quun médecin ou un des administrateurs eut lidée de faire ces comparaisons, voilà le difficile, quoique deux années fussent suffisantes.

Votre article ńMeasurement of characterż a lavantage de

provoquer les réflexions et suscitera peut-être de bonnes observations. Je vous remercie de me lavoir envoyé !

Larticle de Mr Wolff sur les Juifs me paraît un peu trop flatteur à leur égard. Si certaines prescriptions duTalmud sont bonnes pour la santé, il me paraît bien douteux quelles soient suffisamment observées par les fidèles, comme le prétend lauteur. Les injonctions des Juifs et des Musulmans sur les ablutions ne les ont pas rendus généralement propres. Ils sont moins propres que beaucoup de chrétiens auxquels la religion ordonne de mépriser leurs corps, de penser surtout à lâme et à la vie future. Je ne puis croire que la séparation des époux pendant des périodes aussi longues que celles dont parle Mr Wolff, soient réellement observées, surtout les mariages étant précoces, et si on les observe il doit y avoir un palliatif dans une polygamie plus ou moins admise. Les anciens Juifs étaient régulièrement polygames ; ensuite les servantes ont probablement servi de complément à la femme légitime et aujourdhui que se passe-t-il chez les Israélites? Je voudrais dautres témoignages que ceux de Mr Wolff, par exemple le dire de médecins. Il y a beaucoup dactrices juives et de modèles pour les peintres. La prévoyance habituelle chez la race et leurs mariages précoces ont probablement plus dimportance pour régler les murs que les prescriptions de la loi religieuse. Larticle de M. Wolff ma appris du reste bien des choses que jignorais et ses réflexions sont souvent très justes.

Vous avez bien raison de dire que les Juifs sont adaptés à

la vie parasite. Cest une bonne définition des faits. Il faut dire quon les a forcés à cette vie exceptionnelle. Si les difficultés étaient complètement levées pour eux, ils changeraient peut-être. DIsraeli a été un homme détat égal à beaucoup des plus distingués.

Cest un professeur de physiologie juif qui me proposait dexécuter avec moi des expériences sur leffet de lalcoolisme, de la peur, etc. sur les produits dans les lapins, ou les cobayes ou les chiens. Javoue que ces sortes dexpériences me répugnent, mais je crois quelles prouveraient ce que je suppose : que létat momentané des parents influe sur les produits. On la vu maintes fois pour lalcoolisme et je viens de lire dans une revue très sérieuse (Revue dhygiène, Octobre 1884, p, 875) ce qui suit :

ńNotre ami, le Prof. Layet, de Bordeaux, avait à faire (au congrès dhygiène de Lattaie) un rapport sur la restriction volontaire de la natalité du point de vue de ses conséquences humanitaires et socialesy>. Selon lui : ńAu point de vue moral, elle favorise lillégitimité, les 9 dé-partements qui ont les moins de naissances légitimes ont aussi le coefficient dillégitimité le plus élevé... Laccomplissement incomplet dune fonction pervertit les excitations au lieu de les éteindre. Lhabitude de la restriction amène une perturbation du système nerveux des conjoints ; les enfants nés par erreur dans ces conditions se ressentent de la perturbation nerveuse qui a présidé à leur conception. Les aliénés sont plus nombreux dans les départements où les époux ont le moins denfants. ż

ńMr Lunier, de Paris, est convaincu de linfluence de la restriction volontaire sur le développement de létat névrotique, particulièrement de lhystérie et de la névralgie idéo-lombaire chez les femmes, et en cite de nombreuses observations nouvelles. La suppression du droit daînesse fait que le mal sétend comme une tache dhuile dans les campagnes.ż

Certainement dans les pays, comme la France, où la loi assure un partage égal entre les enfants, beaucoup de femmes et de maris ont peur dune 3ème, 4ème ou 5ème grossesse. Leffet de cette peur sur certains enfants est probable. Me voici sur un sujet bien scabreux. Il vaut mieux sarrêter et vous prier, mon cher Monsieur, de me croire toujours votre très dévoué

Alph. de Candolle.

4.23 1885/09/29. Alphonse de Candolle to Francis Galton

Sept. 29, 1885.

My dear sir,

You have kindly sent me your learned article on the inheritance of stature, and I hasten to thank you for it.[33]

Size (or stature) is one of the best characters to study from the point of view of heredity. It is precise and of an

[33] Most likely Galton 1886a.

uncomplicated nature that is suitable for statistics. On the continent we have a quantity of data relating to the size of young men drafted into military service, but nothing on the inheritance of this character. It has often been found that a change in the diet of a population changes the average size. Thus, in the French departments, stature has sometimes increased when corn has been substituted for buckwheat or rye. Inversely, the canton of Bern in Switzerland had a healthy population, strong and of rather high stature, but the peasants began for 50 years to distilll potato brandy. Their wives and even their children drink it. This has resulted in a very visible weakening of the race, and given the number of scrofulous, goitrous, idiots, etc., it is likely that the average size has decreased. We do not have an old document to make it clear, but I will point out to you a recent publication of the Federal Bureau of Statistics which is very interesting for the researches you are dealing with. It is entitled: *Résultats de la visite sanitaire des recrues en automne 1884*,[34] by the Federal Statistical Office, 31 pages in 4°, Bern 1886. Orell Füssli and Co. publishers in Zurich.

This book contains a lot of facts about the size, the thorax, acuity of sight, the profession, and so on, of recruits accepted or refused for military service. The text is in French. It is from the pen of Mr Kummer, head of the Bureau, a very accurate and judicious man.

[34] Results of the Medical Survey of Recruits in the Autumn of 1884.

I have on my table a volume—unfortunately in German—which contains the summary of all that has been collected on the proportion of the sexes and the number of births in various countries, for man and for animals and plants. The title is: *Die Regulierung des Geschlechtverhältnisses bei der Vermehrung der Menschen, Tiere and Pflansen, von Cari Düsing, Dr. Phil. Jena 1884.*

We would assist by translating this book into English or French. However most parts of the text, and of course the tables, are intelligible for people who do not know German well.

The huge proportion of male children born to older fathers struck me markedly. When one of your Lords wants to have sons, you must advise him to marry late! Nutrition, which modifies the proportion of the sexes in the offspring, seems to act differently from what is commonly supposed. On this point the averages obtained for the domestic animals, according to the prior feeding of the parents, are curious and supportive. Veterinarians in Germany have gathered a lot of evidence.

I hope you bring forward in your speeches the many comments you have made. It will be a very interesting exercise.

The publications of the Society of English Anthropology do not arrive here. Would you be kind enough to give me the title and the price of the newspaper of this company? I might find it possible to have it bought by our Reading Society, a sort of literary institution already receiving French and German publications on anthropology.

Receive, my dear sir, the assurance of my very devoted sentiments.

Alph. de Candolle.

29 Sept. 1885.

Mon cher Monsieur,

Vous avez eu la bonté de menvoyer votre savant article sur lhérédité de la stature et je mempresse de vous en remercier.

La taille (soit stature) est bien un des meilleurs caractères à étudier au point de vue de lhérédité. Il est précis et dune nature simple qui sadapte à la statistique. Sur le continent nous avons une quantité de données relatives à la taille des jeunes gens appelés forcément au service militaire, mais rien sur lhérédité de ce caractère. On a constaté souvent quun changement dans lalimentation dune population modifie la taille moyenne. Ainsi, dans les départements français, la stature sest quelquefois élevée quand le maïs a été substitué au sarrasin ou au seigle. Inversément le canton de Berne en Suisse avait une belle population, forte et de taille plutôt élevée, mais les paysans se sont mis depuis 50 ans à distiller de leau de vie de pommes de terre. Leurs femmes et même leurs enfants en boivent. Il en est résulté un affaiblissement très visible de la race, et vu le nombre de scrofuleux, goitreux, idiots etc., il est probable que la taille moyenne a diminué. Nous navons pas de document ancien qui permette de le bien constater, mais je vous signalerai une publication récente du Bureau fédéral de Statistique fort intéressante pour les recherches dont vous vous occupez. Elle est intitulée : Résultats de la visite sanitaire des recrues en automne 1884, par le Bureau de Statistique fédéral, 31 pages in 4°, Berne 1886. Orell Füssli et Cie éditeurs à Zurich.

On trouve dans ce cahier beaucoup de faits sur la taille, le thorax, lacuité de la vue, la profession, etc. des recrues acceptées ou refusées pour le service militaire. Le texte est en français. Il est de la plume de Mr Kummer, chef du Bureau, homme très exact et judicieux.

Jai sur ma table un volume—malheureusement en allemand—qui contient le résumé de tout ce quon a réuni sur la proportion des sexes et le nombre des naissances en divers pays, chez lhomme et dans les animaux et les plantes. Le titre est : Die Regulierung des Geschlechtverhâltnisses bei der Vermehrung der Menschen, Tiere und Pflansen, von Cari Düsing, Dr Phil. Jena 1884.

On rendrait service en traduisant cet ouvrage en anglais ou en français. Cependant plusieurs parties, et naturellement les tableaux sont intelligibles pour les personnes qui ne savent pas bien lallemand.

La proportion énorme des enfants mâles quand le père est âgé ma extrêmement frappé. Quand un de vos Lords voudra avoir des fils, il faut lui conseiller de se marier tard ! La nutrition qui modifie la proportion des sexes dans les produits paraît différente de ce quon suppose communément. Sur ce point les moyennes obtenues pour les animaux domestiques, selon la nourriture préalable des parents, sont curieuses et positives. Les vétérinaires en Allemagne ont réuni beaucoup de faits.

Jespère que vous avancez dans vos relevés des nombreuses observations que vous avez faites. Ce sera un travail bien intéressant.

Les publications de la Société danthropologie anglaise narrivent pas ici. Aurlez-vous la bonté de me donner le titre et le prix du journal de cette société? Je trouverais peut-être moyen de le faire acheter par notre Société de Lecture, sorte dinstitution littéraire qui reçoit déjà des publications françaises et allemandes sur lanthropologie.

Recevez, mon cher Monsieur, lassurance de mes sentiments très dévoués.

Alph. de Candolle.

4.24 1885/10/04. Francis Galton to Alphonse de Candolle

42, Rutland Gate, London.
October 4, 1885.

My Dear Sir,

Excuse delay in reply, as though I date from town I am

still in the country. Let me first cordially thank you for your kind letter and the many interesting remarks it contains.

(1) I have written to the Secretary of the Anthropological to tell you exactly what the annual cost of the journal is, I think it is č1, viz. 4 parts at 5/- each. Also I told him to send for your acceptance from me, a recent number in which there is an exceedingly good paper about the Jews, illustrated by some rather successful "composite" photographs of Jews by myself, which it may amuse you to look at.

(2) I have ordered both the books you speak of: thank you very much for telling me of the latter especially, I mean that about the sex of the child.

(3) You were so kind as to send me some time ago your investigation into the colour of hair, and I feel myself open to blame for not having drawn attention to it already at the Anthropological or elsewhere, but the fact is that I wanted to work up my own data, and to give both results at the same time. My data are now worked up, but there still remains something to be done, so that there will be a little further delay.

Did you ever consider the physiology of clear green eyes—bright green I mean, such as Dante says Beatrice had? The common often repeated statement that blue eyes are merely the effect of seeing pigment through a semi-transparent medium, and that there is only one sort of pigment, cannot possibly explain the existence of blue and green eyes, both equally translucent. There must be a green pigment somewhere. I have asked all our best physiologists, and have looked through many German and French memoirs, thus far in vain, for a rationale.

I am assured that the pigment particles are not so minute as to affect the light by any iridescent effect. In short, that the blue and green cannot be due to such causes as those that make the waters of the Rhone blue, and that of some of the Tyrolese rivers, green.

Believe me,
Very faithfully yours,
Francis Galton.

4.25 1887/05/26. Francis Galton to Alphonse de Candolle

42, Rutland Gate, London.
May 26, 1887.

My Dear Sir,

It gave me great pleasure to receive the "Extrait" from the *Revue d'Anthropologie* of May 15 containing your article on the relative healthfulness of the brown and blond types. You had told me of the suspicion you then had of the accuracy of the American references and I had long wished to see your article. Their statistics are clearly imperfect from neglecting important data. No doubt, however, you have remarked that the soldiers—the accepted men—of German birth are usually ranked high for their physique in Baxter's Statistics (see Vol. I, p. 169 and again pp. 182, 215, on the one side, and pp. 199, 206, 227, on the other).[35] I cannot find in my English statistics any sign of the dark race supplanting the fair. The persistence of the proportions during four generations between them (see Diagram on p. 405, *Royal Soc. Proceed.*, 1886—I send the memoir "Hereditary Eye-Colour" for your acceptance[36]) is very remarkable. Neither do my data show that either is more prolific or less healthy than the other. The data are but scanty; still I imagine that the English climate and surroundings may be equally suited to the two types. Moreover the Scandinavian contingent to our population, contributing largely to the blond type in Eastern England and Scotland, seems the most vigorous though least aesthetic of all our stocks. I have failed in obtaining trustworthy results from my data concerning sexual predilection for, or aversion from, concolour marriages; there are too many interfering causes of importance on which I am insufficiently informed. It is, as you most justly say, among the irregular liaisons that data are most preferably to be sought. Together with the "Hereditary Eye-Colour" I send "Hereditary Stature" which will I fear

[35] Jedediah H. Baxter, *Statistics, medical and anthropological, of the Provost-Marshal-General's Bureau, derived from records of the examination for military service in the armies of the United States during the late war of the rebellion* (Washington, 1875).

[36] Galton 1886b.

hardly interest you being very mathematical in its reasoning, but as the Eye-colour inquiry depends on formulae derived from it I may as well send it also.[37] It also describes my data. Thirdly I send a recent Presidential address, the last part of which beginning at p. 394 may be worth while glancing at.[38]

When I had the great pleasure of making your personal acquaintance a little more than a year ago, you were in domestic anxiety. If you should ever again favour me with a letter, I should be very glad to learn that that anxiety was lessened.[39]

Believe me, my dear Sir,

Very faithfully yours,
Francis Galton.

4.26 1888/04/20. Alphonse de Candolle to Francis Galton

April 20, 1888.

My dear sir,

It is a long time since I should have given you a sign of life in answer to your obliging letter of May 26th, but age has made me very slow and prevents me from doing research of any kind. I would like to know, however, if you are making

[37] Galton 1886c.
[38] Galton 1887.
[39] Candolle's wife Laure Jeanne Victoire died on the 6th of April, 1887.

progress in your useful publications, in which I will always be interested.

One of my last efforts was to write a series of questions about heredity for the Society for Psychological Physiology, founded in Paris. I gathered from Mr. Taine that the Society has received answers and that we are processing them. It remains to be seen to what extent the people questioned were capable of the requisite impartiality and judgment.

Reading the correspondence of your famous cousin Charles Darwin has caused me a lot of pleasure.[40] I would have liked to know the questions he was probing with Sir Joseph Hooker in 1852-54, when I was concerned with the origin of the species from a geographical point of view, which led me in 1855 to note the geological antiquity of the causes of the current distribution. On the other hand, it would have delayed me in research and I would have published my *Géographie botanique raisonnée* only later, after perhaps the classic work of Darwin (1859).[41]

There are in his letters characteristic and apposite phrases about principles and methods in the observational sciences. I want to re-read the three volumes to extract them.

What a remarkable exhibition of the development of ideas

[40] Francis Darwin ed. *The Life and Letters of Charles Darwin including an autobiographical chapter* (London: John Murray, 1887).

[41] That is, if he had pursued inquiries along Darwin's lines it would have delayed his main line of research. Here Candolle regrets a missed opportunity.

of the author! There is nothing like this in Montaigne, Gibbon, Rousseau and others who have written about themselves too late and with an often questionable impartiality.[42]

Winter has caused disasters in Switzerland and elsewhere. Here they are mostly avalanches. Despite this, our lakes and our Alps will always be alluring. Will not you visit them again this summer? It would be very nice for your devoted

Alph. de Candolle.

20 avril 1888.

Mon cher Monsieur,

Il y a longtemps que jaurais dû vous donner signe de vie en réponse à votre lettre obligeante du 26 mai dernier, mais lâge ma rendu très lent et mempêche de faire des recherches daucun genre. Jaimerais pourtant bien savoir si vous avancez dans vos utiles publications, auxquelles je porterai toujours de lintérêt.

Un de mes derniers efforts a été de rédiger pour la Société de Psychologie physiologique fondée à Paris, une série de questions à poser sur lhérédité. Jai su, par Mr Taine, que la Société a reçu des réponses et quon soccupe de les utiliser. Reste à savoir comment les personnes questionnées ont été à la hauteur dimpartialité et de jugement nécessaires.

La lecture de la correspondance de votre célèbre cousin Charles Darwin ma causé beaucoup de plaisir. Jaurais bien

[42] Candolle refers to the (bowdlerized) fragment of autobiography included in the *Life and Letters*.

aimé connaître les questions quil agitait avec Sir Joseph Hooker, en 185254, lorsque je moccupais moi-même de lorigine des espèces, au point de vue géographique, ce qui me conduisait en 1855 à constater lancienneté géologique des causes de la distribution actuelle. Dun autre côté cela maurait retardé dans les recherches et je naurais publié ma Géographie botanique raisonnée que plus tard, après peut-être louvrage classique de Darwin (1859).

Il y a dans ses lettres des phrases caractéristiques et admirables sur les principes et les méthodes dans les sciences dobservation. Je veux relire les trois volumes pour les extraire.

Quelle remarquable exposition des idées successives de lauteur! On ne trouve rien de pareil dans Montaigne, Gibbon, Rousseau et autres qui ont écrit sur eux-même trop tard et avec une impartialité souvent douteuse.

Lhiver a causé des désastres en Suisse comme ailleurs. Ici ce sont surtout des avalanches. Malgré cela nos lacs et nos Alpes auront toujours de lattrait. Ne viendrez vous pas les visiter de nouveau cet été? Ce serait fort agréable pour votre dévoué

Alph. de Candolle.

4.27 1888/05/06. Francis Galton to Alphonse de Candolle

42, Rutland Gate, London.

4.

May 6, 1888.

My Dear Sir,

It gave me very great pleasure to hear from you about a fortnight ago, and I should have replied at once only I thought the enclosed scrap (which might have been printed a week earlier) would interest you and I delayed till I got it. Dr Venn's memoir will not appear till November.[43] He is the author of a most thoughtful book called the Logic of Chance which young statisticians ought to read, for it explains what statistics cannot as well as can do, in a very masterly way.[44] The third edition is just out. If you happened to think of any logically disposed reviewer it would be worth while suggesting this book to him as well deserving notice.

I was very pleased to read how much Charles Darwin valued and profited by your labours and views. What an immensity of work in science has been performed in the last 50 years! It must be an endless pleasure to yourself to look back upon your own large contribution to it. It will be very curious to watch the results obtained from your questions circulated by the *Société de Psychologie physiologique*, and the way in which the veracity of the answers may be tested. I have myself lately had a batch of rather disappointing replies to questions circulated among teachers in schools of all grades,

[43] Possibly Venn 1888.
[44] Venn 1866.

concerning the signs and warnings of mental fatigue.[45] There was great absence of skilful self-analysis and of suggestion, and not a few transparent indications of exaggeration here and of suppression there. I was hearing the other day from a particularly trustworthy source, a list of unveracities of one of our own men of science, formerly one of the leaders of science, but whom I must not indicate further. The general facts and many particulars I had long known, but was surprised to learn how much more there was that I had not known. It is strange that a man who had so little care for truth could succeed in science at all. It is a most painful case of psychological interest and made me think how painfully it would have interested you when writing that paragraph on the general veracity of men of science in your XIXme Siècle.

I had a pleasant summer last year in Eastern Switzerland, etc., but in the autumn fell suddenly ill with a most severe gastric attack at Lugano and was got home somehow in a wagon-lit. Then I fell ill again in another way with violent catarrh, then again in a third way with inflammation of the caecum, and lastly in a fourth way with severe bronchitis. In short I had four separate severe illnesses within five months. I suspect there was some microbic poison at the bottom of it. However I am clear of all illness just now.

I was grieved to see the black-edged paper of your letter, and beg of you to accept my sympathy. I shall deem myself

[45] Galton 1888b.

very fortunate if the next time that I pass through Geneva I shall have the great pleasure of finding you at home and inclined for a half hour's conversation.

Very faithfully yours,
Francis Galton.

4.28 1888/05/28. Alphonse de Candolle to Francis Galton

28 May 1888.

My dear sir,

I regret to learn from your letter of May 8th that you have been sick so long, but fortunately you add that now your health is restored. As for me, the fatigues and sorrows caused by the illness and the death of my wife have singularly weakened my moral faculties, while hearing, sight, and memory are diminished by a natural effect of age.[46] I lost my old energy and my confidence in the possible result of research. I must take my retirement and remember that, having started publishing in 1824, my scientific career did not last less than 64 years. My old taste for statistics still persists, at least when it comes to following good work done by others.

It is therefore with pleasure that I have read your analysis of Dr. Venn's research on the heads of Cambridge students.[47]

[46] Candolle had married Laure Jeanne Victoire Kunkler (1811–1887) in 1832.
[47] Galton 1888a.

He has made many convincing comparisons among young people of the same conditions, ages, etc. who behave differently at the university. For example, compare fearless smokers, mediocre smokers, and non-smokers, both from the point of view of intellectual success and success in bodily exercises. The antagonism between the intellectual and bodily aptitudes so well known to the ancients, would undoubtedly emerge from a statistical comparison in the schools.

About physical exercise I recommend a volume that has just appeared in the international collection of Alcan (formerly Alglave) in Paris. This is Dr. F. Lagrange, *Physiologie des exercices du corps*[48] 1 vol. in 8°, Paris 1888, Price 6 fr. The author deals with the physiology of muscles, nerves, &c. in a very learned and true way, as I judge it, and I noticed a distinction that is not yet mentioned, namely that certain exercises tire both head and body, while others rest the brain while using the muscles. Therefore the first ones (fencing for example) contribute to the overwork that is complained in schools, while the others (walking for example) have no inconvenience and offer a lot of physical benefits. Exercises that require mental application for idlers, stultifying exercises for students who want employment, and mentally fatiguing exercises for clerks, employees, etc., should be recommended.

I knew two distinguished scholars who were not very truthful, but I must say they did not lie about scientific matters,

[48] Physiology of physical exercise.

like their experiments or observations. It was rather to serve a friend, or to hurt someone they did not like. Both loved politics. Men of science sometimes lack moral strength and the result is a willingness to hide certain opinions rather than lie. In general, however, I feel that the habit of research makes for truthfulness. If you go to Geneva this summer I will be very happy to see you. In the heat of the day I may go to the mountains, but it would be neither far nor long.

Receive my dear Sir, the assurance of my most devoted feelings.

Alph. de Candolle.

I will ask for Dr. Venn's book.[49] If it is too mathematical for me I will consign it to some mathematician among my friends.

28 mai 1888.

Mon cher Monsieur,

Je regrette dapprendre par votre lettre du 8 mai que vous avez été si longtemps malade, mais heureusement vous ajoutez que maintenant votre santé est rétablie. Quant à moi les fatigues et le chagrin causés par la maladie et la mort de ma femme ont singulièrement affaibli mes facultés morales pendant que louïe, la vue et la mémoire diminuaient par un effet naturel de lâge. Jai perdu mon ancienne activité et ma confiance dans le résultat possible des recherches. Il

[49] Venn 1866.

faut prendre mon parti de la retraite et me souvenir quayant commencé à publier en 1824, ma carrière scientifique na pas duré moins de 64 ans. Mon ancien goût pour la statistique persiste encore, au moins lorsquil sagit de suivre de bons travaux faits par dautres.

Cest donc avec plaisir que jai lu votre analyse des recherches du Dr Venn sur la tête des étudiants de Cambridge. Il a bien des comparaisons probantes à faire sur des jeunes gens de mêmes conditions, âges, etc. qui se conduisent diversement à luniversité. Par exemple, comparez les fumeurs intrépides, fumeurs médiocres et non-fumeurs, au double point de vue des succès intellectuels et des succès dans les exercices du corps. Lantagonisme entre les aptitudes intellectuelles et corporelles si bien connu des Anciens, ressortirait sans doute dune comparaison statistique dans les écoles.

A propos dexercices je vous recommande un volume qui vient de paraître dans la collection internationale dAlcan (autrefois Alglave) à Paris. Cest Dr F. Lagrange, Physiologie des exercices du corps. 1 vol. in 8°, Paris 1888, Prix 6 fr. Lauteur traite la physiologie des muscles, nerfs, etc. dune manière très savante et vraie, à ce quil me paraît, et jai remarqué une distinction dont on ne parle pas encore, cest que certains exercices fatiguent à la fois la tête et le corps, tandis que dautres reposent le cerveau tout en employant les muscles. Par conséquent les premiers (escrime par exemple) contribuent au surmenage dont on se plaint dans les écoles, tandis que les autres (la marche par exemple) nont aucun inconvénient et offrent beaucoup davantages physiques. Il faut recommander les exercices qui exigent une tension desprit aux oisifs, et les exercices bêtes aux étudiants qui veulent travailler et aux commis, employés, etc. dont la tête est fatiguée.

Jai connu deux savants distingués qui nétaient pas bien véridiques, mais je dois dire quils ne mentaient pas sur des affaires scientifiques, comme leurs expériences ou observations. Cétait plutôt pour rendre service à un ami, ou pour nuire à quelquun quils naimaient pas. Tous deux aimaient la vie politique. Les hommes de science manquent parfois de force morale et il en résulte une disposition à cacher certaines opinions plutôt quà mentir. En général cependant jestime que lhabitude des recherches rend véridique. Si vous passez à Genève cet été je serai très heureux de vous voir. Dans le moment des grandes chaleurs jirai peut-être dans les montagnes, mais ce ne serait ni loin ni pour longtemps.

Recevez mon cher Monsieur, lassurance de mes sentiments les plus dévoués.

Alph. de Candolle.

Je demanderai louvrage du Dr Venn. Sil est trop mathématique pour moi je le communiquerai à quelque calculateur de mes amis.

4.29 1889/09/15. Alphonse de Candolle to Francis Galton

Sept. 15, 1889.

My dear Sir,

I have just read in the *Times* your observations on the Civil Service examinations and I have read also articles in the *Nineteenth Century* on the same subject, which interests me particularly.[50] Here is why: I have a grandson, born in England, graduated from Rugby College, who has been fully occupied for 18 months preparing for a Civil Service exam which (alas!) is never announced and has been postponed a long time. I am aware that the desire is to reduce the number of places and adjourn the examinations because of that. But it would be fair to look at the young people who are prepared. It would be fair play. In France there are exams for admission to polytechnic, military, etc. and exams are held at fixed times each year. So after a few months a young man knows if he is admitted or if he must aim for another career.

Your reflections on the physical conditions to be stipulated in the examinations are perfect when it comes to military service or Indian service, and I believe that in these cases proper measures have already been taken.

As for the Civil Service proper, I notice a capital defect.

[50] Galton 1889b,c.

The same knowledge is required for services of a very different nature. Thus, for sedentary or active occupations, for services of engineer, accountant etc. who ask for mathematics, or diplomatic services, or literary, which require knowledge of languages and history we force the candidates to also stuff their heads with Greek, Latin, mathematics, history, economics, politics, etc. etc. Half of the subjects will not serve a purpose in all careers. It is the Chinese system, in which every scholar is supposed to be fit for anything and useful in any career.

A first desirable reform would be to have two categories of examinations for the young people of the so-called superior division. For some one would require especially literary knowledge, and for the others especially mathematical.

Bodily attributes are not of equal importance in all cases. To inspect public works some muscular strength, height, and good eyesight are advantages. But for the sedentary jobs of the post office, various offices of administration and diplomacy, it is quite irrelevant. People who are a little weak and sedentary by nature are better off in an office than those who are passionate about sports. The most skilful diplomat of the nineteenth century was Talleyrand, who was lame and shortsighted.

The true principle should be to obtain specific men: the right man in the right place. We are moving away from that in the current English Civil Service system. It is perhaps the consequence of the false ideas of the present democracy which judges all men equal and fit for everything, which leads to a general mediocrity. I am happy you continue your research. Your perseverance will be rewarded and already we know how much it is valuable.

Receive my dear Sir, the assurance of my very devoted feelings.

Alph. de Candolle.

PS If you know a job in which it is advantageous to speak and write both French and English, to know enough German and to have extensive knowledge in the historical sciences, I can recommend my grandson, 21 years old.[51] He meets your requirements, of intellectual families through three generations, on the paternal side and as much on the maternal side.

Genève 15 septembre 1889

Mon cher Monsieur

Je viens de lire dans le Times vos observations sur les examens du Civil service et jai lu également des articles dans le Nineteenth century sur le même sujet, qui mintéresse particulièrement. Voici pourquoi: Jai un petit fils, né en Angleterre, sorti avec honneur du collège de Rugby, et qui se fatigue depuis 18 mois à préparer un examen du Civil service, qui (hélas!) nest jamais annoncé et sera peut-être encore renvoyé

[51] Augustin Richard Emile de Candolle (1868–1920). He abandoned studying for the civil service examination but eventually became British Consul to Geneva, after practicing botany there in the family tradition, without formally qualifying in the discipline.

longtemps. Je năžignore pas quon veut réduire le nombre des places et quon ajourne les examens à cause de cela. Mais il serait pourtant équitable davoir égard aux jeunes gens qui se préparent. Ce serait fair play. En France il y a des examens pour lentrée dans les écoles polytechnique, militaire, etc., et des examens qui ont lieu à des époques fixes chaque année. Ainsi au bout de quelques mois un jeune homme sait sil est admis ou sil doit viser à une autre carrière. Vos réflexions sur les conditions physiques à stipuler dans les examens sont de toute justesse quand il sagit du service militaire ou du service Indien, et je crois que dans ces cas on a déjà pris des mesures convenables. Quant au civil service proprement dit je remarque un défaut capital. On exige les mêmes connaissances pour des services dune nature très différente. Ainsi, pour des occupations sédentaires ou actives, pour des services dingénieur, de calculateur etc., qui demandent des mathématiques, ou des services diplomatiques ou littéraires, qui demandent des connaissances de langues vivantes et dhistoire, on oblige les candidats à se bourrer la tête également de grec, latin, mathématiques, histoire, économie politique etc., etc. La moitié des objets ne servira à rien dans chaque carrière, cest le système chinois, dans lequel chaque lettré est supposé apte à tout et utile dans toute carrière. Une première réforme désirable serait davoir deux catégories dexamens pour les jeunes gens de la division dite supérieure. Pour les uns on exigerait des connaissances spécialement littéraires ou pour les autres spécialement mathématiques. Les conditions physiques ne sont pas de même importance dans tous les cas. Pour inspecter des travaux publics une certaine force musculaire, une taille élevée, une bonne vue, sont des avantages. Mais pour des métiers sédentaires de la poste, de divers bureaux dadministration et dans la diplomatie cest assez indifférent. Les gens un peu faibles et sédentaires par nature valent mieux dans un bureau que ceux passionnés de sports. Le plus habile diplomate du XIX e siècle a été Talleyrand, qui était boiteux et myope. Le vrai principe devrait être dobtenir des hommes spéciaux: the right man in the right place. On sen éloigne dans le système actuel anglais du civil service. Cest peut-être la conséquence des idées fausses de la démocratie actuelle qui juge tous les hommes égaux et préparés à tout, ce qui conduit à une médiocrité générale.

Je pense avec plaisir à la continuation de vos recherches. Votre persévérance sera récompensée et déjà on sait à quel point elle est méritoire. Recevez mon cher Monsieur, lassurance de mes sentiments très dévoués.

Alph. de Candolle Court St Pierre 3 Genève

PS. Si vous connaissez un emploi dans lequel il soit avantageux de parler et écrire également bien le français et langlais, de savoir assez lallemand et davoir des connaissances étendues dans les sciences historiques, je pourrai vous recommander mon petit fils âgé de 21 ans. Il rentre dans vos conditions de familles intellectuelles, par trois générations du côté paternel et autant du côté maternel.

4.30 1889/11/13. Francis Galton to Alphonse de Candolle

42, Rutland Gate, London.
November 13, 1889.

My Dear Sir,

The long delay of two months in replying to your very kind letter has been wholly due to the hope that I might have something to say that you would like to hear. The particular scheme about which you wrote, of our introducing marks in our competitive examinations for physical efficiency, has not yet publicly resulted in anything, but from private information I learn officially, though confidentially, that the question will almost certainly be examined into by a very favourably disposed committee of one of our great public Departments, among whose officials the need of high physical efficiency is great.

Also several of our public schools are, I believe, making experiments in marking for it, and in seeing how far the examiners agree between themselves and with the general verdict of those masters who know the boys thoroughly in the cricket field, at football, and in other games.

I venture to send you the paper in full that I read at the British Association (of which the last part was published with good illustrations in Nature, Oct. 31).

On looking at the second page where I have marked a paragraph, you will see how careful I was not to commit the fault you feared in your letter, of supposing that high bodily efficiency is of universal importance. I only speak of professions in which it is.

I was very sorry to hear of the inconvenience to which your grandson has been put, by an absence of an opportunity of competition on which he had reckoned. Probably the expected vacancies did not occur. I do not at all profess to defend the action of our Civil Service Commissioners either in giving a notice of expected examinations which was not fulfilled, or in exacting much the same knowledge from candidates for widely different offices. But they have a very difficult task in fulfilling as far as may be two conflicting wishes. One is not to disturb the regular course of education, so that a youth may be educated at any great school without going to a special "crammer" up to nearly the last moment, and the other is to require a sufficiency of special knowledge.

This is accomplished in some cases by two examinations, the one at a comparatively early age, to qualify for entering; the second one which is special, and not so severe, but that every lad who passed the first might be expected to succeed in the second. Then if he failed in the first, he would be in the same position as other boys who looked forward to any one of a multiplicity of possible careers. No one however seems satisfied with what is now done either in the Government examinations or in the public school teaching; but no one here has yet had the wit to suggest a course that commends itself to the general judgment as an improvement. The question is apparently a most involved one; so many interests and prospects being seriously affected by any change of system.

As regards the particular question you put, as to any satisfactory employment for a person having the high qualifications you mention, clearly they must exist in abundance, but personally I have not any one of them distinctly in view at present. I should have thought that a private secretaryship to some political person would be eminently a post to try for, or that to some person in the higher branches of commerce or manufacture, who has varied foreign connections. All such posts give a young man excellent opportunities for afterwards succeeding by his own efforts, and adequately educated candidates for them are hardly equal in number to the demand.

In concluding let me express the great pleasure that it gave me to receive your kind letter, for there are now few persons whose sympathy I prize more than your own on those many subjects in which we feel a common interest. You say nothing of your health but I trust and believe that it is maintained more fully by far than in the great majority of your contemporaries.

Believe me,
very faithfully yours,
Francis Galton.

4.31 1890/06/23. Alphonse de Candolle to Francis Galton

23 June 1890,

My dear Sir,

I have not thanked you yet for the information you have kindly given me about the Civil Service and the examinations postponed indefinitely. My grandson must have dropped the object after having expended 18 months preparing for it. He went to Germany to learn law.

No longer able to pursue science, I amuse myself observing the decline of my faculties, and for a moment I thought I could add something to your research on the relative influence of Nature and Nurture. I said to myself: Are the faculties that are best maintained at 84 years old, those of birth (nature) or the result of frequent exercise? It turns out that they are both a birth effect and a continual use. Inversely, the faculties that became very weak were weak at the beginning and were hardly cultivated during my life.

Thus I have preserved the faculty of walking better than many old men. But my father had been a great walker in his youth and I always liked walking. I once made steep ascents in the mountains, which is the only exercise I have ever done.

I inherited a weak memory from my mother. Now it is very weak. But I have always had a reluctance to memorize and have always tried to replace memory with notes.

My conclusion is that most men do the things they naturally feel comfortable with, and neglect the ones they are not good at. The use that results from a natural disposition, and the non-use, is also due to a natural weakness. This is contrary to what teachers and many parents think. They want to force young men and women to do what they do not like, while the youth would like to do what they want. There is thus a lot of lost time and energy; but the youth soon escapes the constraint and then we see the young people who are not calculators abandon mathematics, the young ladies close their pianos etc ... The pedagogues want to make all individuals alike, but the individuals prefer to be dissimilar, which will be a greater benefit to society in general.

Have you been informed that naturalists have made great progress over the past two or three years on the process of fertilization in both kingdoms? It is no longer the protoplasm that plays the main role, but the male and female nuclei (nuclei). These nuclei mate. They contain filaments, in a fixed number, whose positions change in a curious manner. You could judge these discoveries by looking at the plates of a memoir of Mr. Guignard, in the Bulletin of the Botanical

Society of France of 1889 which is also in the Proceedings of the Botanical Congress in Paris in 1889. The zoologists observed the same facts.

Accept, I beg you, my dear Sir, the assurance of my most devoted sentiments.

Alph. de Candolle.

23 Juin 1890,

Mon cher Monsieur,

Je ne vous ai pas encore remercié des informations que vous avez bien voulu me donner au sujet du Civil Service et des examens renvoyés indéfiniment. Mon petit-fils a dû ny plus penser, après avoir perdu 18 mois à sy préparer. Il est allé en Allemagne apprendre le droit.

Ne pouvant plus travailler pour la Science, je mamuse à observer le déclin de mes facultés, et jai cru un moment pouvoir ajouter quelque chose à vos recherches sur linfluence relative de Nature and Nurture. Je me disais : Les facultés qui se maintiennent le mieux chez moi à 84 ans, sont-elles celles de naissance (nature) ou le résultat dun exercice fréquent ? Il se trouve quelles sont à la fois un effet de naissance et dun usage continuel. Inversément les facultés devenues très faibles étaient faibles à lorigine et nont guère été cultivées pendant ma vie.

Ainsi jai conservé la faculté de marcher mieux que beaucoup de vieillards. Or mon père avait été un grand marcheur dans sa jeunesse et jai toujours aimé la marche. Jai fait autrefois de fortes marches dans les montagnes, cest le seul exercice que jaie cultivé.

Jai hérité de ma mère une mémoire faible. Maintenant elle est très faible. Or jai toujours eu de la répugnance à apprendre par cur et jai cherché toujours à remplacer la mémoire par des notes.

Ma conclusion est que la plupart des hommes font les choses auxquelles ils se sentent naturellement propres, et négligent celles pour lesquelles ils ne sont pas bien doués. Lusage résulte dune disposition naturelle et le non-usage dune faiblesse aussi de nature. Voilà qui est bien contraire à ce que pensent les instituteurs, les professeurs et beaucoup de parents. Ils veulent forcer les jeunes gens et les jeunes filles à faire ce quils naiment pas, tandis que la jeunesse aimerait faire ce pour quoi chacun se sent bien doué. II y a ainsi beaucoup de temps et de force perdus ; mais la jeunesse échappe bientôt à la contrainte et alors on voit les jeunes gens qui ne sont pas calculateurs abandonner les mathématiques, les jeunes demoiselles fermer leurs pianos etc... Les pédagogues veulent faire tous les individus semblables et les individus voudraient être dissemblables, ce qui serait un grand avantage pour la société en général.

Avez-vous été informé que les naturalistes ont fait depuis deux ou trois ans de grands progrès sur le procédé de la fécondation dans les deux règnes ? Ce nest plus le protoplasma qui joue le principal rôle, mais les noyaux (nuclei) mâles et femelles. Ces noyaux saccouplent. Ils renferment des filaments, en nombre déterminé, dont les positions changent dune manière curieuse. Vous pourriez juger de ces découvertes en regardant les planches dun mémoire de Mr Guignard, dans le Bulletin de la Société Botanique de France de 1889 qui se trouve aussi dans les Actes du congrès de botanique à Paris en 1889. Les zoologistes ont observé les mêmes faits.

Recevez, je vous prie, mon cher Monsieur, lassurance de mes sentiments les plus dévoués.

Alph. de Candolle.

Figure 4.9: Bust of Alphonse de Candolle.

List of Figures

Glossary

Murchison, Sir Roderick 1792–1871. Geologist and frequent president of the Royal Geographical Society. He was also highly influential within professional bodies like the Royal Society. 4

X Club A dining club of Darwinians-without-Darwin, scientists with broadly similar philosophical outlooks distinguished by their promotion of Darwin's heresies. The club was founded on November 3, 1864, but the networks of friendship it reflected existed some time before that. The members were: T.H. Huxley; Joseph Dalton Hooker; Sir John Lubbock; William Spottiswoode; John Tyndall; George Busk; Edward Frankland; Thomas Archer Hirst; and Herbert Spencer.

In its early days the club met at 6PM on the first Thursday of the month in the St. George's Hotel in Albemarle street, just before the meetings of the Royal Society at 8PM, but took the summer off, going on excursions instead (these later fell away). Guests were regularly invited to join the coterie for dinner. After the death of Spottiswoode, meetings were held at the Athenaeum club, which Spottiswoode had never cared to join. The meetings petered out in the 1890s, by which time most of its members had died and the rest had lost interest.

Huxley, a surgeon and paleontologist had known Hooker, a botanist from a famous botanical family, since the early 1850s, through the British Association. They were the prime movers in starting the club, though they later gave the impression that the club was informal and only a way to keep in touch regularly.

The British Association also brought the sociologist, historian and philosopher Herbert Spencer together with Huxley in the early 1850s. Huxley introduced Spencer to the experimental physicist Tyndall. Spencer met the surgeon-naturalist Busk, and through him the banker and ethnographer Lubbock, much later, in the early 1860s. Hooker had already known Lubbock, independently of Spencer, since 1860. Darwin was a neighbour and close friend of Lubbock, who lived just outside the village of Downe. In the 1870s Lubbock became a Whig MP.

The mathematician Hirst and the chemist Frankland were friends of Tyndall; they had all obtained doctorates in Germany in the early 1850s, at the University of Marburg. Frankland and Tyndall were tutored there by Bunsen, while Hirst was supervised by Steggman, and did post-doctoral work under Steiner and Gauss.

The purchase of the *The Reader* in late 1864 was initiated by the first meeting of the X Club, but its members were already strongly associated with the magazine from its inception in 1863. A minority of the new shareholders were club members, but nearly all the club members were shareholders. Galton was a shareholder in the *Reader*, but never a member of the club, as the list was never expanded after the first meeting. He was proposed twice in subsequent years, as was General Richard Strachey, but no action was ever taken. This may have been due to the fact that he was not close to either Huxley or Hooker when the club was first formed, even though he was very close to both Tyndall and Spottiswoode.

The club members proved to have extensive influence through the Royal Society, which they dominated at the upper echelons in the 1870s and early 1880s (though Spencer was never a member there). Aside from being Darwinists, the members were all Liberal Unionists opposed to Home Rule on the grounds of freedom from religion, especially for science. 4

Bibliography

Candolle, Alphonse de (1855). *Géographie botanique raisonée ou exposition des faits principaux et des lois concernant la distribution géographique des plantes de l'époque actuelle.* 2 volumes. Paris: Victor Masson, 1855.

— (1873). *Histoire des sciences et des savants depuis deux siècles.* History of science and scholars for two centuries. Geneva: H. Georg, 1873.

Galton, Francis (1864). *The Knapsack Guide for Travellers in Switzerland.* First edition. Not credited. London: John Murray, 1864.

— (1865). *A Handbook for Travellers in Switzerland.* Not credited. London: John Murray, 1865.

— (1869). *Hereditary Genius. An Inquiry into its Laws and Consequences.* First edition. London: Macmillan, 1869.

— (1872). "On blood-relationship." In: *Proceedings of the Royal Society* 20 (June 13, 1872), pp. 394–402.

— (1873a). "Hereditary Improvement." In: *Fraser's Magazine* 7 (Jan. 1873), pp. 116–30.

Galton, Francis (1873b). "On the Causes which Operate to Create Scientific Men." In: *Fortnightly Review* 13 (Mar. 1, 1873), pp. 345–51.

— (1873c). "Problem 4001: On the extinction of surnames." In: *Educational Times* 26 (Apr. 1873), p. 17.

— (1874a). *English Men of Science*. London: Macmillan, 1874.

— (1874b). "On men of science, their nature and their nurture." In: *Proceedings of the Royal Institution* 7 (Feb. 27, 1874), pp. 227–236.

— (1875a). "A Theory of Heredity." In: *Journal of the Anthropological Institute* 5 (Nov. 9, 1875), pp. 329–348.

— (1875b). "Short notes on Heredity &c. in Twins." In: *Journal of the Anthropological Institute* 5 (Nov. 23, 1875), pp. 324–9.

— (1875c). "The history of twins, as a criterion of the relative powers of nature and nurture." In: *Journal of the Anthropological Institute* 5 (Nov. 23, 1875), pp. 391–406.

— (1878). *Composite portraits made by combining those of many different persons into a single figure*. With Additions. London: William Harrison and Sons, 1878.

— (1879a). *Psychometric Experiments*. London: William Clowes and Sons, 1879.

— (1879b). *Psychometric Facts*. London: Henry King, 1879.

— (1880a). *Visualised Numerals*. London: Harrison and Sons, 1880.

— (1880b). "Visualised numerals." In: *Nature* 21 (Jan. 15, 1880), pp. 252–6.

— (1880c). "Visualised numerals." In: *Nature* 21 (Feb. 5, 1880), p. 323.

— (1880d). "Visualised numerals." In: *Nature* 21 (Mar. 25, 1880), pp. 494–5.

— (1883). *Inquiries into Human Faculty and its Development.* London: Macmillan, Apr. 24, 1883.

— (1886a). "Family likeness in stature." In: *Proceedings of the Royal Society* 40 (Jan. 21, 1886), pp. 42–63.

— (1886b). "Family-likeness in eye-colour." In: *Proceedings of the Royal Society* 40 (May 27, 1886), pp. 402–416.

— (1886c). "Hereditary Stature [President's address]." In: *Journal of the Anthropological Institute* 15 (1886), pp. 488–499.

— (1887). "Address delivered at the anniversary meeting of the Anthropological Institute of Great Britain and Ireland, January 25th 1887." In: *Journal of the Royal Anthropological Institute of Great Britain and Ireland* 16 (Jan. 25, 1887), pp. 387–402.

— (1888a). "On head growth in students at the University of Cambridge." In: *Journal of the Anthropological Institute* 18 (Apr. 24, 1888), pp. 155–156.

— (1888b). "Remarks on Replies by Teachers to Questions Respecting Mental Fatigue." In: *Journal of the Anthropological Institute* 18 (Apr. 10, 1888), pp. 157–68.

Galton, Francis (1889a). *Natural Inheritance.* London: Macmillan, 1889.

— (1889b). "On the advisability of assigning marks for bodily efficiency in the examination of candidates for public services." In: *Report of the British Association for the Advancement of Science* 59 (1889), pp. 471–473.

— (1889c). "The sacrifice of education to examination." In: *Nineteenth Century* 25 (Feb. 1889), pp. 303–8.

— (1908). *Memories of My Life.* London: Methuen, 1908.

Galton, Francis and H. W. Watson (1874). "On the probability of the extinction of families." In: *Journal of the Anthropological Institute* 4 (May 12, 1874). Published a year later in 1875, pp. 138–139.

Morsier, G. de (1972). "Correspondance inédite entre Alphonse de Candolle (1896–1893) et Francis Galton (1822–1911)." In: *Gesnerus: Swiss Journal of the history of medicine and sciences* 29 (1972), pp. 129–160.

Pearson, Karl (1924). *The Life, Letters and Labours of Francis Galton. Volume II: Researches of Middle Life.* Cambridge: Cambridge University Press, 1924.

— (1930). *The Life, Letters and Labours of Francis Galton. Volume IIIb: Characterisation, Especially by Letters; Index.* Cambridge: Cambridge University Press, 1930.

Venn, John (1866). *The Logic of Chance. An Essay on the Foundations and Province of the Theory of Probability, With Especial Reference to Its Logical Bearings and Its*

Application to Moral and Social Science, and to Statistics. Second edition, 1876; Third Edition, 1888. London: Macmillan, 1866.

— (1888). "Cambridge Anthropometry." In: *Journal of the Royal Anthropological Institute* 18 (Apr. 24, 1888), pp. 140–154.

Made in the USA
Monee, IL
04 November 2026

67378733R00125